An Introduction to Dust Explosions

An Introduction to Dust Explosions

Understanding the Myths and Realities of Dust Explosions for a Safer Workplace

Paul Amyotte

AMSTERDAM • BOSTON • HEIDELBERG • LONDON
NEW YORK • OXFORD • PARIS • SAN DIEGO
SAN FRANCISCO • SINGAPORE • SYDNEY • TOKYO

Butterworth-Heinemann is an Imprint of Elsevier

Butterworth-Heinemann is an imprint of Elsevier
The Boulevard, Langford Lane, Kidlington, Oxford, OX5 1GB, UK
225 Wyman Street, Waltham, MA 02451, USA

First published 2013

Notices
Knowledge and best practice in this field are constantly changing. As new research and
experience broaden our understanding, changes in research methods, professional practices,
or medical treatment may become necessary.

Practitioners and researchers must always rely on their own experience and knowledge
in evaluating and using any information, methods, compounds, or experiments described
herein. In using such information or methods they should be mindful of their own safety and
the safety of others, including parties for whom they have a professional responsibility.

To the fullest extent of the law, neither the Publisher nor the authors, contributors, or editors,
assume any liability for any injury and/or damage to persons or property as a matter of
products liability, negligence or otherwise, or from any use or operation of any methods,
products, instructions, or ideas contained in the material herein.

British Library Cataloguing in Publication Data
A catalogue record for this book is available from the British Library

Library of Congress Cataloguing in Publication Data
A catalog record for this book is available from the Library of Congress

ISBN: 978-0-12-397007-7

For information on all Butterworth-Heinemann publications
visit our website at store.elsevier.com

Printed and bound in the United States of America

13 14 15 16   10 9 8 7 6 5 4 3 2 1

Working together
to grow libraries in
developing countries

www.elsevier.com • www.bookaid.org

This book is dedicated to my wife, Peggy, and our children, Jonathan, Matthew, Sarah, and Lauren. I am forever grateful for your unfailing love, support, and understanding.

Contents

Preface   xiii
Author   xv

1. Introduction: Dust Explosions — Myth or Reality?

1.1   Explosion Pentagon   3
1.2   Dust Explosion Myths   4
1.3   Why this Book?   4
1.4   What Do *You* Think?   6
     References   7

2. Myth No. 1 (Fuel): Dust Does Not Explode

2.1   Dust Definition   10
2.2   Determination of Dust Explosibility   12
2.3   An Explosible Non-Explosible Dust   12
2.4   Reality   14
2.5   What Do *You* Think?   15
     References   16

3. Myth No. 2 (Fuel): Dust Explosions Happen Only in Coal Mines and Grain Elevators

3.1   Cyclical Interest in an Ever-Present Problem   18
3.2   Magnitude of the Problem   21
3.3   Reality   26
3.4   What Do *You* Think?   27
     References   28

4. Myth No. 3 (Fuel): A Lot of Dust Is Needed to Have an Explosion

4.1   Guidance from Physics and Chemistry   32
4.2   Practical Guidance   34
4.3   Housekeeping   34
4.4   Reality   36
4.5   What Do *You* Think?   36
     References   38

5.   Myth No. 4 (Fuel): Gas Explosions Are Much Worse Than Dust Explosions

5.1   Hazard and Risk   42
5.2   Example: Likelihood of Occurrence and Prevention   42
5.3   Example: Severity of Consequences and Mitigation   45
5.4   Hybrid Mixtures   45
5.5   Reality   46
5.6   What Do *You* Think?   47
         References   50

6.   Myth No. 5 (Fuel): It's Up to the Testing Lab to Specify Which Particle Size to Test

6.1   Role of Particle Size Distribution   52
6.2   Particle Size Effects on Explosibility Parameters   56
6.3   A Cooperative Endeavor   59
6.4   Reality   60
6.5   What Do *You* Think?   61
         References   62

7.   Myth No. 6 (Fuel/Ignition Source): Any Amount of Suppressant Is Better Than None

7.1   Inerting and Suppression   67
7.2   Minimum Inerting Concentration   69
7.3   Suppressant Enhanced Explosion Parameter   70
7.4   Thermal Inhibitors   71
7.5   Reality   72
7.6   What Do *You* Think?   73
         References   75

8.   Myth No. 7 (Ignition Source): Dusts Ignite Only with a High-Energy Ignition Source

8.1   Industrial Ignition Sources   78
8.2   Standardized Dust Explosibility Testing   79
8.3   Dust Cloud Ignition by Low-Energy Sources   83
8.4   Reality   87
8.5   What Do *You* Think?   88
         References   89

9.   Myth No. 8 (Ignition Source): Only Dust Clouds—Not Dust Layers—Will Ignite

9.1   Dust Layer Ignition   92
9.2   Dust Layer Fires   93
9.3   Reality   95
9.4   What Do *You* Think?   96
         References   99

10. Myth No. 9 (Oxidant): Oxygen Removal Must Be Complete to Be Effective

10.1 Limiting Oxygen Concentration   103
10.2 Candidate Inert Gases   106
10.3 Reality   107
10.4 What Do *You* Think?   108
      References   111

11. Myth No. 10 (Oxidant): Taking Away the Oxygen Makes Things Safe

11.1 Nothing is Safe   114
11.2 Introduction of New Hazards   117
11.3 Management of Change   119
11.4 Reality   120
11.5 What Do *You* Think?   124
      References   124

12. Myth No. 11 (Mixing): There's No Problem If Dust Is Not Visible in the Air

12.1 Primary and Secondary Dust Explosions   128
12.2 Domino Effects   132
12.3 Reality   133
12.4 What Do *You* Think?   135
      References   136

13. Myth No. 12 (Mixing): Once Airborne, a Dust Will Quickly Settle out of Suspension

13.1 Dustiness   140
13.2 Preferential Lifting   143
13.3 Nano-Materials   145
13.4 Flocculent Materials   146
13.5 Reality   149
13.6 What Do *You* Think?   150
      References   153

14. Myth No. 13 (Mixing): Mixing Is Mixing; There Are No Degrees

14.1 Turbulence   156
14.2 Influence of Turbulence   158
14.3 Concentration Gradients   161
14.4 Reality   163
14.5 What Do *You* Think?   164
      References   165

15. Myth No. 14 (Confinement): Venting Is the Only/Best Solution to the Dust Explosion Problem

15.1 Inherently Safer Design 169
15.2 Hierarchy of Controls 171
15.3 Dust Explosion Prevention and Mitigation Measures 172
15.4 Reality 172
15.5 What Do *You* Think? 176
References 179

16. Myth No. 15 (Confinement): Total Confinement Is Required to Have an Explosion

16.1 Degree of Confinement 182
16.2 Explosion Relief Venting 184
16.3 Reality 186
16.4 What Do *You* Think? 186
References 187

17. Myth No. 16 (Confinement): Confinement Means Four Walls, a Roof, and a Floor

17.1 Congestion and Obstacle-Generated Turbulence 190
17.2 Temporary Enclosures 194
17.3 Reality 195
17.4 What Do *You* Think? 195
References 196

18. Myth No. 17 (Pentagon): The Vocabulary of Dust Explosions Is Difficult to Understand

18.1 Dust Explosion Terminology 200
18.2 Gas Explosion Analogies 203
18.3 Right to Know 204
18.4 Reality 204
18.5 What Do *You* Think? 205
References 206

19. Myth No. 18 (Pentagon): Dust Explosion Parameters Are Fundamental Material Properties

19.1 A Quiescent Dust Cloud—The (Nearly) Impossible Dream 208
19.2 The Mystical K_{St} Parameter 210
19.3 Standardized Dust Explosibility Testing (Revisited) 213
19.4 Reality 215
19.5 What Do *You* Think? 215
References 217

20. Myth No. 19 (Pentagon): It Makes Sense to Combine Explosion Parameters in a Single Index

20.1 USBM Indices 220
20.2 Assessment and Management of Dust Explosion Risks 223
20.3 Material Safety Data Sheets 226
20.4 Reality 227
20.5 What Do *You* Think? 227
References 232

21. Myth No. 20 (Pentagon): It Won't Happen to Me

21.1 Safety Culture 236
21.2 Safety Management Systems 238
21.3 Westray Coal Mine Explosion 240
21.4 Reality 242
21.5 What Do *You* Think? 243
References 246

22. Conclusion: Dust Explosion Realities

22.1 Myths and Corresponding Realities 250
22.2 What Do *You* Think? 250
References 254

Index 255

Preface

Thank you for reading this book. I hope you find it helpful in understanding some of the realities of dust explosions as well as some of the myths that abound in this field. This is how the book is structured: an introduction and conclusion, and in between, 20 myths and their corresponding realities. I have chosen the explosion pentagon as a further structural element for my writing because this basic concept has served me well in tracing the origin and propagation of dust explosion incidents. Chapter 1 provides more details on the nature of the book and its intended audience.

My name appears in many of the references given at the end of each chapter; that is simply because mine is the work I know best. There should be no doubt, however, that this research belongs to the many undergraduate students, graduate students, and postdoctoral fellows with whom I have had the pleasure of working over the years. Any errors of interpretation are solely attributable to me.

I have been blessed with the friendship and guidance of numerous colleagues from industry, academia, and government, both in Canada and internationally. Professor Michael Pegg, also of Dalhousie University, introduced me to the world of dust explosions. Dr. Masaharu Nifuku of the National Institute for Advanced Science and Technology, Japan, and Professor Kazimierz Lebecki of the Central Mining Institute, Poland—among many others—opened my eyes to the fact that research excellence and collegiality have no regional or geographical boundaries. I am especially grateful to Dr. Ashok Dastidar of Fauske and Associates, United States, and Professor Faisal Khan of Memorial University, Canada, for their friendship and research partnership.

I am indebted to Professor Piotr Wolanski of the Warsaw University of Technology, Poland, for initiating the series of dust explosion colloquia that has transitioned into the biannual event known as the International Symposium on Hazards, Prevention, and Mitigation of Industrial Explosions (ISHPMIE). The recent 9th ISHPMIE held in Krakow, Poland, provided much helpful material in writing the current book. Similarly, the efforts of the U.S. Chemical Safety and Hazard Investigation Board have been instrumental in developing case studies of industrial dust explosions that have been referenced here.

I also want to acknowledge several individuals who were important members of my team in providing text and visual material for the book. Morgan Worsfold conducted literature searching and identified several key references for analysis. Dr. Meftah Abuswer helped with a critical review of the book once completed in draft form. Jonathan Amyotte obtained high-quality images for many of the figures reproduced from other sources. Finally, Matthew Amyotte created and drew the character appearing at the start of each chapter.

In closing, I wish to acknowledge and thank two people for their special contributions. The first is Professor Rolf Eckhoff of the University of Bergen, Norway. This book would probably not have been written were it not for his suggestion and encouragement. At my request in 2010, Professor Eckhoff reviewed the draft of my keynote conference paper titled "Dust Explosions Happen Because We Believe in Unicorns" (referenced in Chapter 1). He suggested this paper might form the basis for a book on the subject of dust explosion myths and realities. Knowing full well the challenges of preparing a scientific text, he continued to offer words of encouragement throughout the writing process. I am most grateful to Rolf for his kindness in addition to all the technical lessons he has taught me.

My final thank-you is extended to my wife, Peggy. It has been my experience that writing a book requires long periods of uninterrupted time that come at the expense of other activities. After I started this project, her encouragement and support—and most importantly her gift of time—are what enabled me to complete it. I am most grateful to Peggy for all she brings to better my life.

Paul Amyotte
Dalhousie University, Halifax, NS, Canada

Author

Paul Amyotte is a Professor of Chemical Engineering and the C.D. Howe Chair in Engineering at Dalhousie University in Canada. His teaching, research, and practice interests are in the areas of process safety, inherent safety, and dust explosion prevention and mitigation. He has published or presented over 200 papers in the field of industrial safety and is co-author with Professor Trevor Kletz of the second edition of *Process Plants: A Handbook for Inherently Safer Design*. He is also the current editor of the *Journal of Loss Prevention in the Process Industries*.

Introduction: Dust Explosions— Myth or Reality?

Unicorn: a mythical animal generally depicted with…a single horn in the middle of the forehead.

—Merriam-Webster's Online Dictionary

There is a problem, the nature of which is not well understood, in communicating the results of dust explosion testing and research to stakeholders in industry, government, and the public. In a recent article on dust explosions, I was quoted as follows [1] (p. 47):

When I hear about yet another dust explosion, I hang my head. When someone who has been in the industry for a certain number of years says that they didn't know sugar or flour or aluminum could explode because they'd never seen it happen before—that's just wrong.

The answer to this problem is neither as trivial nor as obvious as it may seem. A partial answer—or at least the idea for the paper [2] providing the basis for this book—came from Professor Trevor Kletz during his workshop held as part of the Hazards XXI symposium in Manchester, UK (November 2009). Professor Kletz commented that when poor or impracticable designs are examined, some people may not question the intention of the designers, whereas others may speak up because they see technical oversights and hazards that were not seen before. To illustrate his point, he showed a slide of an animal with what appeared to be a single horn in the center of its forehead (see Figure 1.1). Was it a unicorn? No; the next slide showed the same animal (an oryx) from a different angle, and now it was clear there were two horns on its head (see Figure 1.2). *What we see depends on the way we look* [3].

So perhaps dust explosions do occur, in part, because we believe in unicorns—in myths that lack appropriate elements of the natural, management, and social sciences and engineering principles associated with dust explosion prevention and mitigation. This book explores 20 such myths drawn from my

FIGURE 1.1   Unicorn? *(Photograph courtesy of T. Kletz.)*

FIGURE 1.2 No—oryx! *(Photograph courtesy of T. Kletz.)*

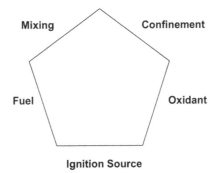

FIGURE 1.3 The explosion pentagon.

research activities and experience in providing dust explosibility test results to industry. Other practitioners and authors would undoubtedly come up with a different set of attitudes and beliefs needing closer examination, but the ones presented here form a useful starting point for a dialogue structured around the explosion pentagon shown in Figure 1.3.

1.1 EXPLOSION PENTAGON

In many respects the explosion pentagon affords us everything we need to know on a fundamental level about dust explosion causation [4]. When the requirements of the pentagon are satisfied, the risk of a dust explosion arises. These requirements include the familiar need for a fuel, an oxidant, and an ignition source, augmented by mixing of the fuel and oxidant, as well as confinement of the resulting mixture. The first of these additional components illustrates a key difference between dust and gas explosions—a solid rather than a gaseous fuel. In a dust/air mixture, the dust particles are strongly influenced by gravity; an essential prerequisite for a dust explosion is therefore the formation of a dust/oxidant suspension. Once combustion of this mixture occurs, confinement (partial or complete) permits an overpressure to develop, thus enabling a fast-burning dust flame to transition to a dust explosion [5].

As helpful as the explosion pentagon may be in understanding why dust explosions occur, it is neutral in terms of how individuals interpret, prevent, and mitigate these requirements. The ensuing chapters illustrate connections between the various pentagon elements and *myth* typified by erroneous belief (the *unicorn*), as well as *reality* expressed through scientific and engineering fact (the *oryx*). The book concludes with a set of 20 facts to counterbalance the 20 myths identified throughout.

1.2 DUST EXPLOSION MYTHS

The myths associated with dust explosions and which are explored in this book are the following [with the applicable pentagon element(s) shown in italics]:

- Dust does not explode (*fuel*).
- Dust explosions happen only in coal mines and grain elevators (*fuel*).
- A lot of dust is needed to have an explosion (*fuel*).
- Gas explosions are much worse than dust explosions (*fuel*).
- It's up to the testing lab to specify which particle size to test (*fuel*).
- Any amount of suppressant is better than none (*fuel/ignition source*).
- Dusts ignite only with a high-energy ignition source (*ignition source*).
- Only dust clouds—not dust layers—will ignite (*ignition source*).
- Oxygen removal must be complete to be effective (*oxidant*).
- Taking away the oxygen makes things safe (*oxidant*).
- There's no problem if dust is not visible in the air (*mixing*).
- Once airborne, a dust will quickly settle out of suspension (*mixing*).
- Mixing is mixing; there are no degrees (*mixing*).
- Venting is the only/best solution to the dust explosion problem (*confinement*).
- Total confinement is required to have an explosion (*confinement*).
- Confinement means four walls, a roof, and a floor (*confinement*).
- The vocabulary of dust explosions is difficult to understand (*pentagon*).
- Dust explosion parameters are fundamental material properties (*pentagon*).
- It makes sense to combine explosion parameters in a single index (*pentagon*).
- It won't happen to me (*pentagon*).

1.3 WHY THIS BOOK?

The answer to the question "Why this book?" is obtained by examining what the current book is intended to be, and what it is not intended to be. Starting with the latter point, this book has not been written as a comprehensive treatise on all important aspects of dust explosions. Such an endeavor would require a different focus in reviewing the intensive research on dust explosions that has been conducted in the public and private sectors over the past several decades. This research has led to many advances including improved understanding of dust explosion fundamentals [6], enhanced mitigation techniques such as venting and suppression [7], and recognition of the role of inherently safer design in dust explosion prevention and mitigation (Chapter 13 in Kletz and Amyotte [8]).

The preceding text references are available to readers who desire advanced treatment in the areas indicated. Additionally, various archival journal articles have been written for specialists in industrial loss prevention and dust explosion research. As explained by Amyotte and Eckhoff [5], recent reviews cover in detail case histories, causes, consequences, and control of dust explosions [9]; the role of powder science and technology in understanding dust explosion phenomena [10]; and the status of developments in basic knowledge and practical applications with respect to dust explosion prevention and mitigation [11].

While the scope of the current book is also related to dust explosion phenomena (specifically the prevention and mitigation of dust explosions), it differs from the other works cited in the preceding paragraph in terms of both motivation and objective. Having said this, I would be remiss in not acknowledging the role played by these resources in writing my own manuscript (as evidenced by the numerous references to them throughout).

The writing of this book has been motivated in equal measure by a desire to aid in the protection of people, business assets, operational production, and the natural environment, and a need to address important communication issues with respect to understanding dust explosions.

More generally, one of the process safety research topics identified in the recently published *Process Safety Research Agenda for the 21st Century* [12] is easy-to-implement process safety methods for industry. Quoting from this document [12] (p. 42):

Due to the sophistication needed to make progress, the gap in the level of theoretical knowledge between academia and most industry experts tends to widen and become an obstacle to communication. This can cause a decrease both in the flow of industry experience to academia and the implementation of newly acquired knowledge to industry. Special effort should be made to counter this trend. Easy-to-implement methods require the developer to fully master the method and the knowledge it is based on in order to describe complex phenomena in simple terms and make the method transparent and user friendly.

These motivational points have led to the objective of exploring the myths and realities associated with dust explosion risk reduction. To achieve this objective, I have attempted to provide extensively referenced facts on dust explosions in a manner that clearly and unambiguously refutes several misconceptions about dust explosions. A key feature in this regard is the closing section of each chapter in which readers are invited to express their own thoughts on questions related to the specific content of the chapter.

1.4 WHAT DO *YOU* THINK?

As noted on its website (www.csb.gov), the U.S. Chemical Safety and Hazard Investigation Board (Chemical Safety Board or CSB) is an independent, non-regulatory federal agency that conducts root cause investigations of chemical accidents at fixed industrial facilities. The reports of its investigations are available on the CSB website for downloading and are often accompanied by video footage and animation of the incident sequence [8]. These reports—incident investigations, case studies, safety bulletins, and urgent recommendations—are an excellent resource for training exercises aimed at learning lessons from previous incidents.

The following excerpt is taken from the CSB document describing the results of a recent investigation effort [13] (p. 2):

This case study examines multiple iron dust flash fires and a hydrogen explosion at the Hoeganaes facility in Gallatin, TN. The first iron dust flash fire incident killed two workers and the second injured an employee. The third incident, a hydrogen explosion and resulting iron dust flash fires, claimed three lives and injured two other workers.

This particular Hoeganaes plant manufactures atomized iron powder for the production of metal parts in the automotive and other industries. Hydrogen is used in the plant's continuous annealing furnaces to prevent oxidation of the iron powder. Further details, including the answers to the following questions, can be found in the CSB case study [13].

Before reading the full report, however, consider the following questions based on your current knowledge and understanding of the explosion pentagon and its various elements:

- *Fuel:* Can metal dusts such as iron explode? What range of iron dust particle sizes would be expected to support an explosion?
- *Ignition Source:* What is a typical energy required to ignite a cloud of iron dust particles with diameters < 75 μm? What about the temperature of a hot surface required to ignite such a dust cloud?
- *Oxidant:* Is it practical to eliminate oxygen-containing air from all plant areas in which a dust explosion might occur?
- *Mixing:* How could iron dust deposits such as those shown in Figure 1.4 be raised into suspension?
- *Confinement:* What would be required for an iron dust flash fire to transition to an explosion?

FIGURE 1.4   Iron dust deposits on elevated surfaces at the Hoeganaes Corporation facility in Gallatin, TN, on February 3, 2011 [13].

REFERENCES

[1] Kenter P. Big bang theory. OHS Canada 2009;25:42–7.

[2] Amyotte PR. Dust explosions happen because we believe in unicorns. Keynote Lecture, Proceedings of 13th Annual Symposium, Mary Kay O'Connor Process Safety Center, Texas A&M University, College Station, TX; October 26–28, 2010. pp. 3–30.

[3] Kletz T. Equipment and procedures that cannot do what we want them to do. Workshop Notes and Slides, Hazards XXI, Institution of Chemical Engineers, Manchester, UK; November 9, 2009.

[4] Amyotte PR. Facing the pentagon. Industrial Fire Journal. First Quarter 2010;34–5.

[5] Amyotte PR, Eckhoff RK. Dust explosion causation, prevention and mitigation: an overview. Journal of Chemical Health & Safety 2010;17:15–28.

[6] Eckhoff RK. Dust explosions in the process industries, 3rd ed. Boston, MA: Gulf Professional Publishing/Elsevier; 2003.

[7] Barton J, editor. Dust explosion prevention and protection. A practical guide. Rugby, UK: Institution of Chemical Engineers; 2002.

[8] Kletz T, Amyotte P. Process plants. A handbook for inherently safer design, 2nd ed. Boca Raton, FL: CRC Press, Taylor & Francis Group; 2010.

[9] Abbasi T, Abbasi SA. Dust explosions—cases, causes, consequences, and control. Journal of Hazardous Materials 2007;140:7–44.

[10] Eckhoff RK. Understanding dust explosions. The role of powder science and technology. Journal of Loss Prevention in the Process Industries 2009;22:105–16.

[11] Eckhoff RK. Dust explosion prevention and mitigation. Status and developments in basic knowledge and in practical application. International Journal of Chemical Engineering 2009. Article ID 569825, 12 pp.

[12] MKOPSC. A frontiers of research workshop. Process safety research agenda for the 21st century. A policy document developed by a representation of the global process safety academia (October 21–22, 2011). College Station, TX: Mary Kay O'Connor Process Safety Center, Texas A&M University System; 2012.

[13] CSB. Case study—Hoeganaes Corporation: Gallatin, TN—metal dust flash fires and hydrogen explosion. Report No. 2011-4-I-TN. Washington, DC: U.S. Chemical Safety and Hazard Investigation Board; 2011.

Myth No. 1 (Fuel): Dust Does Not Explode

In this chapter we explore the notion that dust does not—or cannot—explode. Possible reasons for the existence of this myth are identified, and suggestions for refuting it on the basis of observable facts are given.

No blame or ulterior motives are ascribed to holding the belief that dusts do not explode; this comment applies equally to all the myths described in this book. My experience has been, however, that many people are surprised to learn that dust explosions can and do occur. It may be a friend who is amused to hear that I have spent the better part of my professional career working in the field of dust explosion prevention and mitigation. Or it may be a client who was previously unaware that the material we had just tested for the client company did in fact explode and could cause significant harm should it do so in an industrial setting.

I recall one such client who, after learning that the powder his employees were handling was explosible, remarked that I had just sent the company down a very expensive path of installing explosion prevention and protection measures. This seemed a curious comment given that it was not me who had decided to process the particular material, and it was certainly not me who had made the material explosible in the first place. I was reminded of Richard Feynman's closing passage in his minority report on the causes of the *Challenger* space shuttle disaster [1] (p. 169), with underlining added here for emphasis:

For a successful technology, reality must take precedence over public relations, for <u>nature cannot be fooled</u>.

Perhaps it is the word itself—*dust*—that creates the perception of particulate matter being non-explosible. In a household context, dust is indeed likely to be seen more as a nuisance from the perspectives of cleanliness and allergy avoidance. The word *gas*, on the other hand, may impart a more general sense of hazard awareness because of its use in relation to applications such as cooking (stoves) and motor vehicles (internal combustion engines).

It does appear there is a stronger linkage between the words *gas* and *explosible* than between the solid counterpart *dust* and *explosible*. Note that in this book I have used *explosible* to denote a substance capable of exploding, but also one which is not intended to explode under industrial conditions. This is to distinguish such materials from an *explosive* such as dynamite which is intended to explode under controlled conditions to achieve a desired objective (e.g., rock blasting for roadway or tunnel construction).

2.1 DUST DEFINITION

Morgan and Supine [2] note the distinction made by the U.S. National Fire Protection Association between a *dust* and a *combustible dust*. NFPA 68 [3] defines a dust as any finely divided solid, 420 µm or 0.017 in. or less in diameter (i.e., material capable of passing through a U.S. No. 40 Standard Sieve). A combustible dust is defined as a combustible particulate solid that presents a fire or deflagration hazard when suspended in air or some other oxidizing medium over a range of concentrations, regardless of particle size or shape [3].

So essentially, one should not view a defined boundary of 420 μm as a sharp delineation between dusts that are explosible and those that are non-explosible. What determines whether a given particulate material represents a dust explosion hazard are its actual chemical composition (Section 2.2) in addition to physical parameters such as particle size (Chapter 6) and particle shape. As noted by Febo [4] (p. 62):

…simply looking at a dust is not necessarily a reliable way to identify a combustible dust hazard.

For example, with respect to particle shape and its effect on explosibility, Figures 2.1 and 2.2 show scanning electron microscope (SEM) images of samples of polyethylene. Given the discussion to follow in Chapter 3, these samples must be considered (at least initially) to be explosible based solely on their composition. But what of the particle size? Again, based on the discussion in Chapter 6 of the current book, because both samples pass through a 200-mesh sieve (nominally < 75 μm), they are certainly fine enough to support combustion.

Concerning particle shape, the fibrous (or flocculent) nature of the sample shown in Figure 2.2 should not be used to make the assessment that it is non-explosible simply because it does not look like the sample shown in Figure 2.1 (which is composed of near-spherical particles). Such a conclusion could have disastrous consequences given that both polyethylene samples were determined to be explosible and capable of generating explosion overpressures of approximately 7 bar(g) (i.e., seven times atmospheric pressure) and rates of pressure rise as high as 300–400 bar/s in a volume of 20 L [5].

S-4700 3.0kV 12.2mm x80 SE(U)   500um

FIGURE 2.1   SEM micrograph of –200 mesh spherical polyethylene sample [5].

FIGURE 2.2   SEM micrograph of −200 mesh fibrous polyethylene sample [5].

2.2 DETERMINATION OF DUST EXPLOSIBILITY

One implication of the definitions given in Section 2.1 is that all dusts are not necessarily combustible. As described by Eckhoff [6], dust explosions generally arise from the reaction of a fuel with oxygen to generate oxides and heat. Thus, dust explosions cannot occur with materials that are already stable oxides, such as silicates and carbonates [6]. This explains why limestone (calcium carbonate, $CaCO_3$) finds use as an explosion inertant in coal mines. Limestone acts as a heat sink when subjected to high temperatures and may eventually decompose into CaO and CO_2; however, it will not explode [7].

But are there materials that will explode when in particulate form and when the explosion pentagon criteria are satisfied? The answer as we know from industrial experience is, of course, yes. (See Chapter 3 for numerous examples.)

If doubt exists as to the explosible nature of a given substance, the material should be tested (Chapters 18 and 19) using standardized equipment and procedures such as those developed by the American Society for Testing and Materials (e.g., ASTM [8]). While literature tabulations (e.g., NFPA [3] and Eckhoff [6]) and online databases (e.g., IFA [9]) of dust explosion parameters can be helpful as indicators of explosibility, such sources cannot be seen as a substitute for actual test data on the material in question.

2.3 AN EXPLOSIBLE NON-EXPLOSIBLE DUST

As a case in point, consider the issue of whether fly ash is explosible. Conventional wisdom might hold that fly ash does not explode because ash is material left over *after* combustion. If, however, combustion is incomplete and a waste fly ash stream becomes contaminated with unburned fuel, a non-explosible

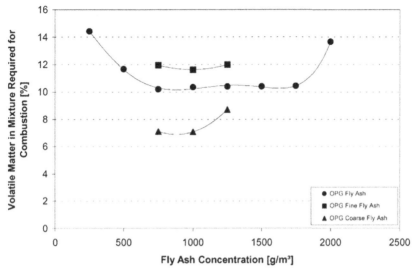

FIGURE 2.3   Volatile percentage required for an explosible fly ash mixture [10].

fly ash can transition to an explosible mixture. This scenario is known as *fuel carryover* and has been documented for pulverized coal utility plants [10].

Figure 2.3 illustrates the explosibility boundaries determined for mixtures of Ontario Power Generation (OPG) fly ash and Pittsburgh pulverized coal (mass mean diameter of 60 μm). Volatile contents of 7%, 10%, and 12% by weight were sufficient to generate explosible mixtures for fly ash having mass mean diameters of 14, 39, and 106 μm, respectively. These results reinforce the earlier comment that chemical composition of a dust is a key factor in determining its explosibility. Further discussion of this point, in particular the role of volatile content for organic dusts, is given in Chapter 5.

The practical implications of the data in Figure 2.3 are displayed by Figure 2.4, which shows the burned coveralls worn by a process plant operator involved in a coke dust/fly ash explosion and ensuing fire. The incident occurred when an air hose was used to clear a plugged hopper containing the coke dust/fly ash mixture; the resulting dust cloud then encountered a gas-fired heater and exploded. Structural damage was minimal (presumably due to limited confinement), although the operator received non-life-threatening burns.

The coke dust had been previously tested and assessed as non-explosible; operating procedures were based on this premise. The introduction of higher volatile content fly ash, however, rendered the mixture explosible as demonstrated by the incident itself. One outcome of the company's investigation was to update the material safety data sheet (or MSDS) for the coke dust/fly ash mixture to indicate that the dust explosion hazard was more severe than for the individual components alone.

FIGURE 2.4 Burned coveralls from coke dust/fly ash explosion and fire (Anonymous, 2009. Personal communication, with permission).

2.4 REALITY

Dusts do explode; not all, but certainly more than enough to warrant a cautious approach. To do otherwise would be to perpetuate a myth, not unlike the one now associated with the Buncefield incident in the UK—i.e., cold petrol (gasoline) does not explode. As noted by Kletz [11] (p. 2):

The underlying cause of the explosion at Buncefield was that all the people and organizations involved in design, operations and maintenance believed that cold petrol vapour had never exploded in the open air. They were unaware that such explosions had occurred in Newark, New Jersey in 1983, Naples, Italy in 1995, St Herblain, France in 1991 and elsewhere.

In a similar vein, Marmo et al. [12] describe an explosion and fire that killed four workers at a plant processing raw pomace oil to obtain edible olive oil. According to the authors, olive oil is well known as non-flammable, and raw pomace oil is generally not considered to be flammable. In this particular incident, however, some of the batches of pomace oil were later found to have a lower than normal flash point (as low as 18.5 °C) because of a higher than usual hexane content [12]. Once again we see that it is the chemical composition of a substance that determines its hazardous characteristics—not the name by which it is called. We are also reminded of Feynman's comment on the intractable nature of *nature* itself [1].

A relevant dust explosion case study in this regard is the incident that occurred at Malden Mills Industries in Methuen, MA, in 1995. As described by Frank [13], the Malden Mills facility manufactured various textile products including one incorporating nylon fibers or flock with a high length-to-diameter ratio. Explosion of this material in an area where high-pressure hoses were being used to clean equipment [13], and the resulting fire, injured 37 employees [14].

The following excerpt from the extensive review of dust explosion incidents conducted by Taveau [14] (p. 43) is noteworthy:

Managers and employees did not understand that the fibers were an explosion hazard before the disaster.

Taveau [14] further comments on how the Malden Mills incident demonstrates the critical importance of dust explosion hazard identification. This is excellent advice that reminds us to always take the necessary steps to determine with the greatest certainty possible whether or not a given dust is explosible.

2.5 WHAT DO *YOU* THINK?

Figure 2.5 shows an SEM image of a polyamide 6.6 (nylon) sample with a nominal dtex of 3.3 and length of 0.5 mm. Dtex or decitex is a unit of measure for the linear density of fibers; it is equivalent to the mass in grams per 10,000 meters of a single filament. For the nylon sample shown in Figure 2.5, a dtex of 3.3 corresponds to a fiber diameter of 19 μm.

Would you consider this sample to be a *dust* according to the definition given in Section 2.1? Answering this question essentially amounts to reconciling a defined diameter of 420 μm with an actual fiber diameter of 19 μm and length of 500 μm (i.e., 0.5 mm). So perhaps a more relevant question is: For the nylon flock shown in Figure 2.5, and from a dust explosion prevention and mitigation perspective, which is the more important definition to consider—*dust* or *combustible dust*?

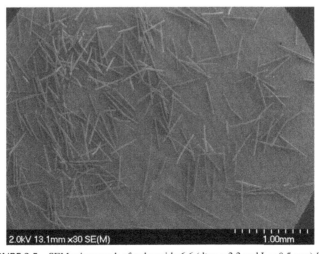

2.0kV 13.1mm x30 SE(M)   1.00mm

FIGURE 2.5   SEM micrograph of polyamide 6.6 (dtex = 3.3 and L = 0.5 mm) [15].

What evidence do you have that nylon flock can be an explosion hazard? Recall the Malden Mills incident described briefly in Section 2.4. You may also wish to consult the case study presented by Marmo [16] or perhaps go to the IFA database [9] (Section 2.2) and search for *nylon flock*.

How would you undertake to determine whether or not the sample in Figure 2.5 would explode as a dust cloud? Note that this question refers to a specific lot of nylon, not nylon flock in general. (Section 2.2 will be helpful here, as will the paper by Iarossi et al. [15])

REFERENCES

[1] Feynman RP. The pleasure of finding things out. The best short works of Richard P. Feynman, Robbins J, editor. Cambridge, MA: Helix Books/Perseus Books; 1999.

[2] Morgan L, Supine T. Five ways new explosion venting requirements for dust collectors affect you. Powder and Bulk Engineering 2008;22(July):42–9.

[3] NFPA. NFPA 68, Standard on explosion protection by deflagration venting. Quincy, MA: National Fire Protection Association; 2007.

[4] Febo HL. Combustible dust hazard recognition—an insurer's view. Bruges, Belgium: Proceedings of 13th International Symposium on Loss Prevention and Safety Promotion in the Process Industries, vol. 2; 2010, (June 6–9, 2010), pp. 59–66.

[5] Amyotte PR, Cloney CT, Khan FI, Ripley RC. Dust explosion risk moderation for flocculent dusts. Journal of Loss Prevention in the Process Industries 2012;25:862–9.

[6] Eckhoff RK. Dust explosions in the process industries, 3rd ed. Boston, MA: Gulf Professional Publishing/Elsevier; 2003.

[7] Amyotte PR. Solid inertants and their use in dust explosion prevention and mitigation. Journal of Loss Prevention in the Process Industries 2006;19:161–73.

[8] ASTM. ASTM E1226–10, Standard test method for explosibility of dust clouds. West Conshohocken, PA: American Society for Testing and Materials; 2010.

[9] IFA. GESTIS-DUST-EX, Database: combustion and explosion characteristics of dusts. Sankt Augustin, Germany: Institut für Arbeitsschutz der Deutschen Gesetzlichen Unfallversicherung; 2012. Available at: http://www.dguv.de/ifa/en/gestis/expl/index.jsp; (last accessed June 6, 2012).

[10] Dastidar AG, Amyotte PR. Explosibility boundaries for fly ash/pulverized fuel mixtures. Journal of Hazardous Materials 2002;92:115–26.

[11] Kletz T. Lessons Forgotten/TCE Comment. The Chemical Engineer 2010;827:2.

[12] Marmo L, Piccinini N, Russo G, Russo P, Munaro L. Multiple tank explosion of pomace oil reservoirs. Yokohama, Japan: Proceedings of Eighth International Symposium on Hazards, Prevention, and Mitigation of Industrial Explosions; September 5–10, 2010.

[13] Frank WL. Dust explosion prevention and the critical importance of housekeeping. Process Safety Progress 2004;23:175–84.

[14] Taveau J. Secondary dust explosions: how to prevent them or mitigate their effects? Process Safety Progress 2012;31:36–50.

[15] Iarossi I, Amyotte PR, Khan FI, Marmo L, Dastidar AG, Eckhoff RK. Explosibility of polyamide and polyester fibers. Krakow, Poland: Proceedings of Ninth International Symposium on Hazards, Prevention, and Mitigation of Industrial Explosions; July 22–27, 2012.

[16] Marmo L. Case study of a nylon fibre explosion: an example of explosion risk in a textile plant. Journal of Loss Prevention in the Process Industries 2010;23:106–11.

Chapter 3

Myth No. 2 (Fuel): Dust Explosions Happen Only in Coal Mines and Grain Elevators

Having established in the preceding chapter that dusts can and do explode, here we examine the belief that dust explosions occur only (or primarily) in coal mines and grain elevators. Let us begin with a look back in history.

One of the first recorded accounts of a dust explosion was written in 1795 by Count Morozzo, who gave a detailed description of an explosion in a flour warehouse in Turin, Italy [1]. Although the explosion pentagon was not known as a causal framework at the time, it is interesting to note the pentagon elements in the following passage from the Count's report [2] (p. 50):

The boy, who was employed, in the lower chamber, in collecting flour to supply the bolter below, dug about the sides of the opening, in order to make the flour fall from the upper chamber into that in which he was; and, as he was digging, rather deeply, a sudden fall of a great quantity took place, followed by a thick cloud, which immediately caught fire, from the lamp hanging to the wall, and caused the violent explosion here treated of.

Fifty years later in 1845, Michael Faraday and a coworker elucidated the key role of coal dust in the devastating explosion in the Haswell (UK) coal mine the previous year [3]. The significance of this finding lies in the fact that coal dust had now been shown to explode in the absence of firedamp (beyond the gas accumulation typically required for the initial ignition sequence). Prior to this observation, firedamp (methane) had been believed to be solely responsible for all such mine explosions.

To this day, dust explosions are often thought of as events that occur predominantly in underground coal mines and industrial grain processing facilities. This perception is accentuated by mine explosions throughout the world (especially, it seems, in China—whether or not they involve coal dust), as well as incidents such as the 1998 DeBruce grain elevator explosion in the United States [4]. (See Figure 3.1.)

There is no denying the visceral impact of images such as that shown in Figure 3.1, especially when coupled with the knowledge that the DeBruce incident resulted in seven fatalities and injuries to three employees, five contractors, and two visitors [4]. In a similar vein, the potential for devastating consequences resulting from a coal dust explosion is graphically illustrated by Figure 3.2. This sequence of photographs shows the fireball caused by a methane-triggered coal dust explosion emerging from a mine portal. (These pictures were taken by the late Kenneth L. Cashdollar of the former U.S. Bureau of Mines and then the National Institute for Occupational Safety and Health in the United Sates. Ken's photography skills were second only to his expertise as a consummate dust explosion researcher and analyst.)

3.1 CYCLICAL INTEREST IN AN EVER-PRESENT PROBLEM

Interest in the dust explosion problem is cyclical. At an international specialist meeting held in Montreal, Canada, in 1981, Professor Bill Kauffman of the University of Michigan commented [5] (p. 305):

FIGURE 3.1   Damage on west side of the headhouse—DeBruce grain elevator explosion (Haysville, KS) [4].

There seems to be little disagreement that the genesis for the revival of interest in agricultural dust explosions was the series of explosions which occurred during the Christmas Season of 1977 in four U.S. grain elevators.

Kauffman [5] gives an accounting of the human and financial toll of these incidents as shown in Table 3.1.

It can similarly be argued that the current impetus for renewed concern about dust explosions (at least in North America) is a series of high-profile incidents that occurred in the United States over the previous decade and were the subject of investigation by the U.S. Chemical Safety Board (CSB). None of these incidents involved coal dust or grain dust, but rather polyethylene (Figure 3.3) [6], phenolic resin (Figure 3.4) [7], aluminum (Figure 3.5) [8], and sugar (Figure 3.6) [9]. The most unfortunate and devastating aspect of these (see Table 3.2) and other industrial dust explosions [10] is that they have all caused significant loss of life and injury.

Just as the dust explosion problem is not limited to coal or grain dust, neither is it limited to those materials involved in the incidents shown in Figures 3.3–3.6.

FIGURE 3.2   Methane-triggered coal dust explosion—Bruceton Experimental Mine (Pittsburgh, PA). *(Photographs courtesy of K.L. Cashdollar).*

TABLE 3.1 Human and Financial Impacts of Dust Explosions in U.S. Grain Elevators during December 1977

Date	Location	Fatalities	Injuries	Damages (Time of incident)
December 21, 1977	Wayne City, IL	1	0	$1.5 million
December 22, 1977	Westwego, LA	36	10	$30.0 million
December 22, 1977	Tupelo, MS	4	15	$1.0 million
December 27, 1977	Galveston, TX	18	22	$25.0 million

(data from Kauffman [5])

FIGURE 3.3 Polyethylene dust explosion—West Pharmaceutical Services (Kinston, NC) [6].

3.2 MAGNITUDE OF THE PROBLEM

As reported by Amyotte and Eckhoff [11], Frank [12] further illustrates the wide scope of the dust explosion problem by using incident data reported by the U.S. CSB and FM Global to show that dust explosions have occurred,

FIGURE 3.4   Phenolic resin dust explosion—CTA Acoustics (Corbin, KY) [7].

FIGURE 3.5   Aluminum dust explosion—Hayes Lemmerz International–Huntington (Huntington, IN) [8].

for example, in the following industries with the indicated typical commodities:

- Wood and paper products (dusts from sawing, cutting, grinding, etc.)
- Grain and foodstuffs (grain dust, flour)
- Metal and metal products (metal dusts)
- Power generation (pulverized coal, peat, and wood)
- Rubber
- Chemical process industry (acetate flake, pharmaceuticals, dyes, pesticides)
- Plastic/polymer production and processing

FIGURE 3.6   Sugar dust explosion—Imperial Sugar Company (Port Wentworth, GA) [9].

TABLE 3.2 Human Impact of Selected Dust Explosions in the United States during the First Decade of the 21st century

Dust	Year	Location	Fatalities	Injuries
Polyethylene	2003	Kinston, NC	6	38
Phenolic Resin	2003	Corbin, KY	7	37
Aluminum	2003	Huntington, IN	1	6
Sugar	2008	Port Wentworth, GA	14	36

(data from CSB [6–9])

- Mining (coal, sulfide ores, sulfur)
- Textile manufacturing (linen flax, cotton, wool).

Additional (and corroborating) evidence of the wide-ranging and insidious nature of the hazards of combustible dusts is given by Figure 3.7. In viewing these data, the question may arise as to their origin—i.e., why the United States and why Germany? The answer lies, at least partially, in the requirements for incident reporting and record keeping in these countries. It should be noted, however, that dust explosions are as prone to under-reporting as other process-related incidents, especially if plant personnel are not impacted or the event falls in the category of a near-miss.

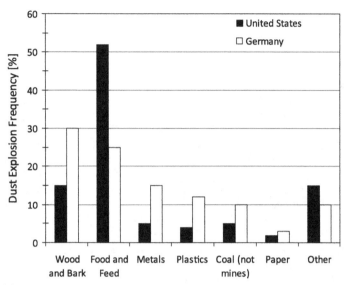

FIGURE 3.7 Frequency of dust explosions for various materials involved in 1,120 incidents in the United States and Germany during the period 1900–1956 [13].

At this point, mention must be made of another pioneer of dust explosion research—the late Wolfgang Bartknecht. There is a geographical connection in that Dr. Bartknecht was originally from Germany and enjoyed a most successful career with Ciba-Geigy in Basel, Switzerland [14] (on the border between northern Switzerland and southern Germany). One of his many lasting contributions is his seminal treatise of dust explosions [15] that articulates in a practical manner the results of extensive testing at the laboratory- and industrial-scale. It is no coincidence that significant advances in our understanding of dust explosion phenomena have been achieved by German and Swiss-German researchers and practitioners.

Data more recent than the period covered by Figure 3.7 are given for the United States in the study on combustible dust hazards conducted by the CSB [10]. The CSB researched the period 1980–2005 and identified 281 major combustible dust incidents resulting in 119 fatalities, 718 injuries, and significant facility damage [10]. Material categories used in the analysis were (i) food, (ii) wood, (iii) metal, (iv) plastic, (v) coal, (vi) inorganic, and (vii) other [10]. Industry classifications were (i) food products, (ii) lumber and wood products, (iii) chemical manufacturing, (iv) primary metal industries, (v) rubber and plastic products, (vi) electric services, (vii) fabricated metal products, (viii) equipment manufacturing, (ix) furniture and fixtures, and (x) other [10].

The discussion to this point has been focused on materials and, by extension, industries susceptible to dust explosions. But what of the typical process units

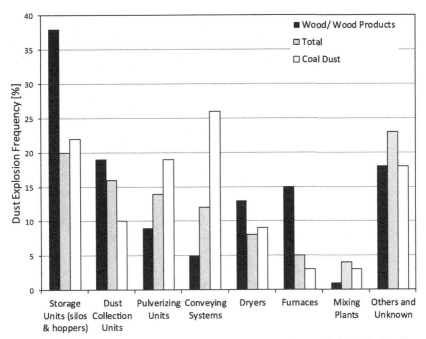

FIGURE 3.8 Frequency of dust explosions for various process units used in industries handling wood and coal dusts in Germany [13].

involved? Abbasi and Abbasi [13] present the following combined list of operations and specific equipment in which dusts are generated or handled: (i) size reduction, (ii) conveying, (iii) pneumatic separation, (iv) screening and classification, (v) mixing and blending, (vi) storage, (vii) packing, (viii) settling chambers, (ix) cyclones, (x) filters, (xi) scrubbers, (xii) electrostatic precipitators, (xiii) driers (tray, rotary, fluidized-bed, pneumatic, and spray), and (xiv) dust-fired heaters.

Figure 3.8 presents data (again, involving Germany) on the frequency of dust explosions occurring in a number of different units used in the processing of wood and coal dusts. Here we see evidence of the attention required in the use of, for example, silos, hoppers, and pulverizing units. These data are generally consistent with those determined in the CSB study [10] covering a broad range of industries as previously described; the CSB work identified silos, hoppers, mixers, and grinders as being involved in numerous dust explosions [10]. Additionally, the CSB cautions that the most prevalent equipment involved in the incidents reviewed were dust collectors [10] (e.g., electrostatic precipitators as given in the general listing of Abbasi and Abbasi [13]).

As an example of a pulverizing unit, Figure 3.9 shows a hammermill used in a wood-processing facility to accomplish size reduction of sawdust and wood chips [16]. Magnetic devices to remove tramp metal (a potential

FIGURE 3.9   Hammermill with top cover open [16].

ignition source), explosion relief vents (Chapter 15), and administrative controls (Chapter 15) restricting access to enclosures housing hammermills are all employed in an attempt to reduce the risk of explosion to an acceptable level. Yet this risk remains high because of the very nature of the functioning of the device; all that is needed in normal operation to complete the explosion pentagon is an ignition source. Hence, hammermills are designed and built strong enough to withstand the overpressure resulting from a dust explosion originating inside the unit.

3.3 REALITY

Dust explosions do not occur only in coal mines and grain elevators. They can arise in any scenario in which combustible dusts are stored, transported, processed, or otherwise handled. It is not only coal dust and grain dust that should be of concern to managers, operators, and other workers in an industrial facility handling bulk powders. Loss of life and injury, damage to facility assets, business interruption, and harm to the natural environment can result from dust explosions in a wide range of industrial applications.

Previous discussion in this chapter has provided ample evidence for these factual statements. If further proof is desired, reference can be made to the authoritative text by Eckhoff [1]. The appendix section of this text gives dust explosibility data organized in the following categories:

- Cotton, Wood, Peat (e.g., cellulose, paper dust, peat of various moisture contents)
- Food, Feed (e.g., dextrose, coffee, rice flour)
- Coal, Coal Products (e.g., activated carbon, brown coal, bituminous coal)
- Other Natural Organic Products (e.g., wheat gluten, oil shale dust, grass dust)

TABLE 3.3 Analysis of 269 Dust Explosions Occurring in Japan during the Period 1952–1995

Dust type	Number of incidents	Fatalities	Injuries
Cellulosic Materials	28	7	84
Chemical Synthetic Materials	36	12	79
Coal	13	7	41
Food and Feed	46	17	109
Inorganic Materials	31	9	28
Intermediate Additives	50	13	69
Metals	64	41	153
Miscellaneous Materials	1	0	4

(adapted from Abbasi and Abbasi [13] with original data from Nifuku et al. [17])

- Plastics, Resins, Rubber (e.g., epoxy resin, polypropylene, polyurethane)
- Pharmaceuticals, Cosmetics, Pesticides (e.g., ascorbic acid, methionine, paracetamol)
- Intermediate Products, Auxiliary Materials (e.g., benzoic acid, pectin, viscose flock)
- Other Technical/Chemical Products (e.g., organic dyestuff, soap, wax raw material)
- Metals, Alloys (e.g., bronze powder, silicon, zinc)
- Other Inorganic Products (e.g., petroleum coke, sulfur, titanium monoxide)
- Other Materials (e.g., ash concentrate, toner, zinc stearate/bentonite).

Using categories similar to those in the preceding list, Table 3.3 provides still further evidence of the potentially life-altering impacts of dust explosions caused by a wide range of materials.

3.4 WHAT DO *YOU* THINK?

Because you are reading this book, it seems reasonable to conclude that you may have one or more dusts involved in your facility operations. Here are a few questions to consider, consistent with the material in this chapter:

- What are those dusts?
- Are they combustible?
- How are they being processed (e.g., is drying or size reduction involved)?
- What process units are involved (e.g., mixers, conveyors, etc.)?
- Are dust collectors used in your facility?

These questions are an attempt to distinguish between the terms *hazard* and *risk*—as in, many materials are explosible (*hazard*), so how are they being handled, transported, processed, etc. (*risk* in terms of explosion likelihood and severity of the consequences)? This topic is treated in greater detail in Chapter 5.

As a note of caution, simply because a particular dust or process unit is not mentioned in the current chapter does not mean it can be eliminated as a hazard or risk factor. In the case of particulate material, recall the discussion in Sections 2.2 and 2.5 concerning the usefulness of dust explosibility tabulated data and databases, as well as the necessity of standardized testing.

Finally, even if you are not in the "dust-handling business," you can still participate in this exercise. Consider a common food product—say, powdered milk. Is this material explosible? What process unit would seem essential to the production of powdered (i.e., solid) milk from liquid milk? (Internet searches, both generally and of selected databases, can be very helpful.)

REFERENCES

[1] Eckhoff RK. Dust explosions in the process industries, 3rd ed. Boston, MA: Gulf Professional Publishing/Elsevier; 2003.

[2] Morozzo C. Account of a violent explosion which happened in the flour-warehouse, at Turin, December the 14th, 1785; to which are added some observations on spontaneous inflammations. From: The Memoirs of the Academy of Sciences of Turin (London: The Repertory of Arts and Manufactures) [Version published with foreword by N. Piccinini, Politecnico di Torino (1996)]. 1795

[3] Eckhoff RK. Understanding dust explosions. The role of powder science and technology. Journal of Loss Prevention in the Process Industries 2009;22:105–16.

[4] Kauffman CW. The DeBruce grain elevator explosion. St. Petersburg, Russia: Proceedings of Seventh International Symposium on Hazards, Prevention, and Mitigation of Industrial Explosions, vol. III. July 7–11, 2008. pp. 3–26.

[5] Kauffman CW. Agricultural dust explosions in grain handling facilities. In: Fuel-air explosions. Lee JHS, Guirao CM, editors. Waterloo, ON: University of Waterloo Press; 1982 [Proceedings of the International Conference on Fuel-Air Explosions held at McGill University, Montreal, Canada, November 4–6, 1981]. pp. 305–47.

[6] CSB. Investigation report—dust explosion—West Pharmaceutical Services, Inc. Report No. 2003-07-I-NC. Washington, DC: U.S. Chemical Safety and Hazard Investigation Board; 2004.

[7] CSB. Investigation report—combustible dust fire and explosions—CTA Acoustics, Inc. Report No. 2003-09-I-KY. Washington, DC: U.S. Chemical Safety and Hazard Investigation Board; 2005.

[8] CSB. Investigation report—aluminum dust explosion—Hayes Lemmerz International–Huntington, Inc. Report No. 2004-01-I-IN. Washington, DC: U.S. Chemical Safety and Hazard Investigation Board; 2005.

[9] CSB. Investigation report—sugar dust explosion and fire—Imperial Sugar Company. Report No. 2008-05-I-GA. Washington, DC: U.S. Chemical Safety and Hazard Investigation Board; 2009.

[10] CSB. Investigation report—combustible dust hazard study. Report No. 2006-H-1. Washington, DC: U.S. Chemical Safety and Hazard Investigation Board; 2006.

[11] Amyotte PR, Eckhoff RK. Dust explosion causation, prevention and mitigation: an overview. Journal of Chemical Health & Safety 2010;17:15–28.

[12] Frank WL. Dust explosion prevention and the critical importance of housekeeping. Process Safety Progress 2004;23:175–84.

[13] Abbasi T, Abbasi SA. Dust explosions—cases, causes, consequences, and control. Journal of Hazardous Materials 2007;140:7–44.

[14] Schaerli A. Obituary. In memory of Wolfgang Bartknecht (1925–2005). Journal of Loss Prevention in the Process Industries 2005;18:v–vi.

[15] Bartknecht W. Dust explosions. Course, prevention, protection. Berlin: Springer-Verlag; 1989.

[16] Amyotte PR, Pegg MJ, Khan FI, Nifuku M, Yingxin T. Moderation of dust explosions. Journal of Loss Prevention in the Process Industries 2007;20:675–87.

[17] Nifuku M, Matsuda T, Enomoto H. Recent developments of standardization of testing methods for dust explosion in Japan. Journal of Loss Prevention in the Process Industries 2000;13:243–51.

Chapter 4

Myth No. 3 (Fuel): A Lot of Dust Is Needed to Have an Explosion

FIGURE 4.1   Damage to portal at No. 1 main, Westray coal mine [1].

How much combustible dust, then, is needed for an explosion to occur? Here we address this question from the perspective of layered dust raised into suspension to form a dust cloud that will ignite given a sufficiently energetic ignition source. So what is the required thickness of such a dust layer? We will first look at a specific example.

On May 9, 1992, a methane-triggered coal dust explosion killed 26 miners at the Westray mine located in Plymouth, Nova Scotia, Canada. An indication of the destructive overpressures generated underground can be seen in Figure 4.1, which shows surface damage at the mine site. The subsequent public inquiry reported that one of the contributing factors was the presence of coal dust layers several centimeters thick throughout the mine workings [1]. (See also Chapter 21 for a discussion of root causes.)

On a personal note, the Westray explosion remains one of the most important events in my professional engineering career. It occurred about 150 km from where I have lived and worked for most of my life—literally in my own backyard. This devastating tragedy changed the lives of hundreds, perhaps thousands, of people in the province of Nova Scotia and elsewhere. For months during 1992, local newspapers reported the grim tale of the Westray mine explosion; I recall articles from that time giving reports of miners trudging through coal dust deposits a foot thick in some places. It is, therefore, neither unusual nor surprising (although erroneous) for people to conclude that dust explosions require deep layers and correspondingly huge amounts of combustible material.

4.1 GUIDANCE FROM PHYSICS AND CHEMISTRY

In fact, the amount of layered coal dust at Westray that could be dispersed by an aerodynamic disturbance and then combusted with the available oxygen would have been much less than the reported thicknesses. Similar considerations apply to many other dust explosion incidents including the Imperial Sugar Company

FIGURE 4.2 Left: Sugar dust accumulation on steel belt drive motor in silo tunnel (October 2006). Right: Cornstarch accumulation (spill) under cornstarch silo (March 2008) [2].

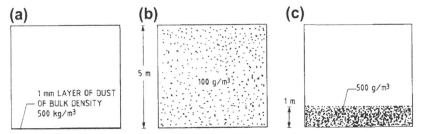

FIGURE 4.3 Illustration of the potential for dust cloud generation from a dust layer [3].

explosion [2], which involved significant accumulations of granulated and powdered sugar. (See Figure 4.2.)

The text by Eckhoff [3] and the resource article by Amyotte and Eckhoff [4] further illustrate in a general manner the importance of dust layer thickness. Figure 4.3 shows the implications of the equation

$$C = (\rho_{\text{bulk}})(h/H) \tag{4.1}$$

where ρ_{bulk} is the bulk density of a dust layer, h is the layer thickness, H is the height of the dust cloud produced from the layer, and C is the resulting dust concentration.

From Equation 4.1 and as shown in Figure 4.3a and b, a 1-mm-thick layer of a dust of bulk density 500 kg/m³ on the floor of a 5-m high room will generate a cloud of average concentration 100 g/m³ if dispersed evenly all over the room. Such a concentration is of the order of the minimum amount required to initiate an explosion for many dusts (termed the minimum explosible concentration, or MEC). Figure 4.3c shows that partial dispersion up to 1 m above the floor yields a dust concentration of 500 g/m³; this is a concentration that is of the order of the optimum concentration (i.e., the concentration producing the most devastating overpressures and rates of pressure rise) for many combustible dusts.

4.2 PRACTICAL GUIDANCE

These observations help to explain the advice given by experienced industrial practitioners on the matter of acceptable combustible dust layer thicknesses. Their comments, although anecdotal, have a firm foundation in the physics and chemistry of dust explosions. Scientific underpinning by the aforementioned difficulties in physically dispersing and chemically reacting excessively thick dust deposits is intrinsic to the following expressions:

- There's too much layered dust if you can see your initials written in the dust [1].
- There's too much layered dust if you can see your footprints in the dust (Anonymous, 1996. Personal communication, with permission).
- There's too much dust if you can't tell the color of the surface beneath the layer (Freeman, R., 2010. Personal communication, with permission).
- I tell my plant managers to write their name on their business card and then place the card on a surface known to collect dust. It's time to clean up when they can't read their name because of layered dust (Anonymous, 2012. Personal communication, with permission).

4.3 HOUSEKEEPING

The removal of dust deposits by good housekeeping practices is a primary line of defense against dust explosions. (See also Chapter 15.) In his recent examination of the U.S. National Fire Protection Association (NFPA) criteria for dust layer thicknesses, Rodgers [5] reports on several possible misinterpretations of such housekeeping requirements—one of which is the notion that only layers thicker than 1/32 in. (0.8 mm) present a hazard. He comments that while there would of course be a layer thin enough that it cannot support a dust cloud explosion, this thickness would depend on several factors related to the initiating event (strength, proximity, and orientation to the layered surface) as well as the aerodynamic features of the dust itself [5].

In addition to the treatment of housekeeping given by Rodgers [5], the articles by Frank [6] and Frank and Holcomb [7] provide helpful information on the topic. In particular, Frank and Holcomb [7] offer the following advice based on NFPA concepts to effectively address the housekeeping challenge (in the desired order of application):

- Design and maintenance of equipment to contain dust so that it does not escape and does not have to be cleaned up
- Dust capture at release points
- Use of physical barriers to limit the extent of dust migration and the size of room to be cleaned
- Facility design for easy and effective cleaning
- Establishing and enforcing housekeeping schedules

- Ensuring that housekeeping programs address all areas in which dust might accumulate
- Ensuring that housekeeping activities are performed safely.

These items are in general accordance with a fundamental construct known as the *hierarchy of controls*, which is covered in detail in Section 15.2. For example, the first measure listed here specifically targets the inherent safety concept of hazard elimination; in the words of Frank and Holcomb [7]:

The easiest/most effective housekeeping is the housekeeping you do not need to do.

The last two points in the preceding list describe the dual requirements of cleaning all dust-laden surfaces and doing so in a fashion that minimizes explosion risk. This is consistent with another of the possible housekeeping misinterpretations given by Rodgers [5]—that only dust accumulations on the floor require consideration. As a specific example, the U.S. Chemical Safety Board has reported that in the case of the Imperial Sugar Company explosion, routine housekeeping involved the use of compressed air to remove layers of accumulated dust from packaging equipment [2]. There are at least two concerns with this procedure; first, it would generate a dust cloud during the actual procedure, meaning that the housekeeping activity itself was not being conducted safely. Figure 4.4 shows an alternative dust removal method, in this case for the collection of samples to determine dust accumulation rates.

Second, the practice would raise dust into suspension to the point where it could accumulate on elevated horizontal surfaces [2]. This situation has been shown previously in Figure 1.4 for iron dust deposits at the Hoeganaes facility in Gallatin, TN, and the damaging effects of this layered dust raining down and

FIGURE 4.4   Use of a portable vacuum for dust collection from round ductwork to enable measurement of dust accumulation rates [7]. *(Photograph courtesy of Kimberly-Clark Corporation.)*

igniting have been described by the CSB [8]. A particularly insidious hazard is the presence of combustible dust layers in locations (elevated or otherwise) that are not readily visible. As described in Chapter 12, the polyethylene explosion at West Pharmaceutical Services in Kinston, NC [9] (Section 3.1 and Figure 3.3) offers a valuable lesson in this regard.

4.4 REALITY

The amount of layered dust which, once airborne, would be required to sustain a dust explosion is often grossly overestimated. Even seemingly harmless dust layers have the potential to rapidly escalate the risk of a dust explosion. Good housekeeping practices leading to the removal of dust deposits are critical; the scale of concern needs to be focused on the millimeter or less level.

The absurdity of accepting thick dust layers as the operational norm can be seen in the following example. Coal dust in underground mines is typically inerted with rock or stone dust (e.g., limestone or dolomite); see Section 2.2 and Chapter 7. Because rock dust is a thermal inhibitor, large amounts are usually required to prevent the explosion pressure from rising to destructive values; depending on a number of factors such as coal particle size and volatile content, the required mass of rock dust may be four or five times that of the coal dust [10].

Consider a work area covered in a foot-thick layer of coal dust (as per the Westray mine). Approximating the bulk densities of the coal dust and rock dust as equal, and assuming an inerting ratio (ratio of rock dust to coal dust) of four, then this one-foot layer of coal dust must be mixed with a four-foot layer of rock dust resulting in a five-foot deep wall of non-explosible mine dust. Imagine working in this environment and imagine the vast quantities of rock dust the mine would have to purchase—absurd, indeed.

The preceding example is obviously an extreme one and is intended more to make a point than to suggest it would actually happen. One might argue that if housekeeping was so obviously inadequate, why would rock dusting practices not be similarly poor? (This is a valid argument as demonstrated in Chapter 21.) So perhaps it is a rhetorical question, but it is still one that needs to be asked: *Why not remove as much of the initial coal dust deposit as possible and then treat the remaining thin layer of coal dust with a sufficient amount of rock dust?*

4.5 WHAT DO *YOU* THINK?

Here are some questions to consider based on the subject matter of dust layers and housekeeping requirements:

- Are there regulatory or standards-driven requirements for dust layer thicknesses in your industry?
- Are there hidden spaces in your facility in which dust could collect and remain out of sight until an external disturbance lofts it into suspension?

- What housekeeping procedures have you developed at your facility to remove combustible dust? Have you adopted a hierarchical arrangement as recommended in Section 4.3? Do you avoid the use of compressed air to "clean up" dust?
- What would be the results of applying Equation 4.1 to the unique specifications of your facility? You could consider this from a couple of perspectives. First, using the bulk density and the minimum explosible concentration (as C, the resulting dust concentration) of your dust, and choosing a realistic value of the height of an enclosed space, calculate the layer thickness required to achieve the MEC (minimum explosible concentration). Or, using the bulk density of your dust and an actual measured value of layer thickness, and choosing a realistic value of the height of an enclosed space, calculate the resulting dust concentration and compare it to the MEC value of your dust.

Relatively simple expressions like Equation 4.1 cannot account for all parameters such as those mentioned earlier in Section 4.3 (strength, proximity, and orientation to the layered surface of the initiating event, as well as the aerodynamic features of the dust itself [5]). They can, however, be quite helpful in removing complexities that are not needed to understand the interplay among important variables. For example, Equation 4.1 clearly shows that for fixed values of layer thickness and cloud height, the resulting dust concentration is directly proportional to the bulk density of the powder. Different dusts will have different values of bulk density, so the application of Equation 4.1 is therefore material-specific.

Consider the data shown in Table 4.1. Note that the layer thickness, h, has been fixed at 0.8 mm (1/32 in.) and the cloud height, H, at 5 m. Representative values of bulk density, ρ_{bulk}, were taken from an online bulk density chart [11]. As an exercise, verify the values of the resulting dust cloud concentration, C, by means of Equation 4.1. Now compare these values of C with the following representative values of minimum explosible concentration (MEC): calcium stearate (30 g/m^3), coal dust (80 g/m^3), corn flour (60 g/m^3), and iron powder (300 g/m^3). Note that these values of MEC are for illustrative purposes only and do

TABLE 4.1 Material-Specific Application of Equation 4.1

Material	ρ_{bulk} [g/cm^3]	h [mm]	H [m]	C [g/m^3]
Calcium stearate	0.32	0.8	5	50
Coal dust	0.56	0.8	5	90
Corn flour	0.82	0.8	5	130
Iron powder	2.80	0.8	5	450

not refer to any particular particle size distribution or dust composition (beyond the generic identifiers given).

Is there a general conclusion that can be made from the preceding exercise?

REFERENCES

[1] Richard KP, Justice. The Westray story—a predictable path to disaster. Report of the Westray Mine Public Inquiry Halifax, NS, Canada: Province of Nova Scotia; 1997.

[2] CSB. Investigation report—sugar dust explosion and fire—Imperial Sugar Company. Report No. 2008-05-I-GA Washington, DC: U.S. Chemical Safety and Hazard Investigation Board; 2009.

[3] Eckhoff RK. Dust explosions in the process industries, 3rd ed. Boston, MA: Gulf Professional Publishing/Elsevier; 2003.

[4] Amyotte PR, Eckhoff RK. Dust explosion causation, prevention and mitigation: an overview. Journal of Chemical Health & Safety 2010;17:15–28.

[5] Rodgers S. Application of the NFPA 654 dust layer thickness criteria—recognizing the hazard. Process Safety Progress 2012;31:24–35.

[6] Frank WL. Dust explosion prevention and the critical importance of housekeeping. Process Safety Progress 2004;23:175–84.

[7] Frank WL, Holcomb ML. Housekeeping solutions. Proceedings of Symposium on Dust Explosion Hazard Recognition and Control: New Strategies Baltimore, MD: The Fire Protection Research Foundation; May 13–14, 2009.

[8] CSB. Case study—Hoeganaes Corporation: Gallatin, TN—metal dust flash fires and hydrogen explosion. Report No. 2011-4-I-TN Washington, DC: U.S. Chemical Safety and Hazard Investigation Board; 2011.

[9] CSB. Investigation report—dust explosion—West Pharmaceutical Services, Inc. Report No. 2003-07-I-NC Washington, DC: U.S. Chemical Safety and Hazard Investigation Board; 2004.

[10] Amyotte PR, Mintz KJ, Pegg MJ. Effect of rock dust particle size on suppression of coal dust explosions. Transactions of the Institution of Chemical Engineers. Part B (Process Safety and Environmental Protection) 1995;73:89–100.

[11] Bulk Density Chart. http://www.powderandbulk.com/resources/bulk_density/ material_bulk_density_chart _c.htm; 2012 (last accessed September 8, 2012).

Chapter 5

Myth No. 4 (Fuel): Gas Explosions Are Much Worse Than Dust Explosions

Broad generalizations on the differences and similarities between gas and dust explosions are a bit like the proverbial comparison of "apples with oranges." While flammable gases and combustible dusts are both *hazards* (as apples and oranges are both fruit), it is more advantageous from a loss prevention perspective to examine each fuel type within the context of *risk* (i.e., the likelihood of an explosion occurring and the severity of its consequences). This is the subject of the current chapter.

First, though, let us consider the origin of the notion that gas explosions are much worse than dust explosions—"much worse" being a subjective qualifier that could mean the holder of this opinion believes gases are much easier to ignite than dusts, they produce much higher explosion overpressures than dusts, or they are just generally more of a problem in industry than dusts. We begin our examination with a return to the explosion pentagon of Chapter 1.

The requirements for an oxidant, an ignition source, and some degree of confinement are arguably comparable for gases and dusts in the sense that these pentagon elements could be the same for each fuel type—for example, normal atmospheric conditions with an open flame in an enclosed space. (Although the term *unconfined vapor cloud explosion* has been used in the past, it should be recognized that some degree of confinement is still provided in such scenarios by the interface between the cloud and its surroundings.)

Key differences for gases and dusts arise in the fuel and mixing pentagon elements. As described in Chapter 1, the first of these additional components means that with dust one is dealing with a solid rather than a gaseous fuel. In combination then, the fuel/oxidant system is by definition heterogeneous (multiphase) for dusts and homogeneous (single-phase) for gases. As also explained in Chapter 1, since dust particles are strongly influenced by gravity, an essential prerequisite for a dust explosion is the formation of a dust/oxidant suspension via adequate mixing. While this means there must be some level of turbulence in a dust cloud, a gaseous fuel/oxidant mixture can be achieved through diffusional mixing, particularly with lighter-than-air gases (e.g., methane). Thus, a quiescent flammable gas/oxidant mixture can be achieved in which the smallest entities of fuel and oxidant are separated only by molecular distances; the same cannot be said for a combustible dust/oxidant mixture (at least within the confines of the earth's gravitational field).

So the heterogeneous nature of a dust/air mixture results in additional steps (e.g., dust lifting) related to the physics of dust explosions. The existence of more than one phase also means there are more chemistry-related steps involved in the explosion process for a particulate fuel than for a gaseous fuel. Together, these points can create the misconception that dust explosions are thus slow to occur and unfold as compared to gas explosions.

Comprehensive treatments of the dust explosion problem such as the text by Eckhoff [1] may be consulted for detailed descriptions of dust explosion reaction mechanisms. Here, it is enough to comment that while some dusts such as fine carbon powder do undergo heterogeneous combustion (i.e., reaction of solid carbon with gaseous oxygen), many dust explosions actually occur as gas explosions once sufficient volatile matter has been generated from the solid material. For example, plastics such as polyethylene would first melt and vaporize to yield a gaseous fuel that undergoes homogeneous combustion with the available oxygen. The step-wise dust combustion process may involve the additional complication of a solid, outer oxide layer as in the case of some metals such as aluminum.

Cashdollar [2] has provided visual evidence for organic dusts, which typically explode in a sequence involving heating to the point of pyrolysis followed by ignition of the evolved volatile matter and subsequent gas-phase flame propagation. Figure 5.1 shows scanning electron microscope (SEM) images of Pittsburgh seam bituminous coal dust before and after explosion in a laboratory-scale chamber. The burned char residue is seen to consist of rounded particles (or cenospheres), some of which have become fractured or display "blow holes" from which volatiles have been emitted [2]. It is these volatiles that burn in a coal dust explosion leading to increased temperature and pressure in an enclosed volume.

As mentioned at the outset of this chapter, the hazard presented by flammable gases and combustible dusts manifests itself in the form of explosion risk. We now turn to an examination of these important process safety concepts.

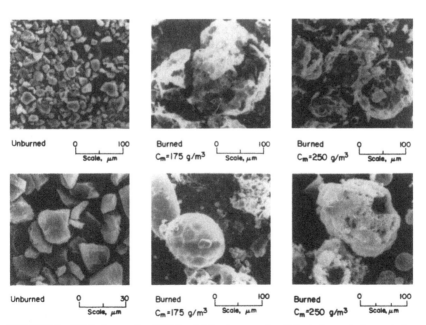

FIGURE 5.1   SEM micrographs of Pittsburgh bituminous coal particles before and after explosion [2]. C_m is the dust cloud concentration at the time of explosion.

5.1 HAZARD AND RISK

While the terms *hazard* and *risk* are often used interchangeably, they are not the same entity and their definitions are quite different [3]:

> *Hazard:* The potential of a machine, equipment, process, material, or physical factor in the working environment to cause harm to people, environment, assets, or production.
>
> *Risk:* The possibility of injury, loss, or environmental incident created by a hazard. The significance of risk is a function of the *probability* (or *likelihood*) of an unwanted incident and the *severity* of its consequences.

A natural consequence of the preceding definitions is that risks are determined by assessment of the likely consequences of identified hazards. Thorough hazard identification is, therefore, key to the effective management of risk; one cannot manage the risk arising from a hazard that has not been identified. With respect to dust explosions and as seen previously in the current book, it is essential to determine whether a given material actually constitutes an explosion hazard, and if so, the degree of hazard as represented by various explosibility parameters. Armed with this knowledge, one can address the issue of risk reduction by means of effective prevention and mitigation techniques. Expressed from a practical, industrial perspective, hazard analysis identifies *what can go wrong*, whereas risk analysis deals with the *probable consequences* of these events in terms of probable loss of life; probable injuries; and probable loss of property, production capacity, and market shares [4].

At this point, it will be helpful to formally present and define the explosibility parameters referred to in the preceding paragraph. Some have already been introduced in this book during discussion of explosion overpressures, rates of pressure rise, and minimum explosible concentrations. Table 5.1 gives a listing of these and other important dust explosibility parameters. The first three entries in Table 5.1 [P_{max}, $(dP/dt)_{max}$, and K_{St}] relate hazard to explosion consequence severity while the remainder are associated with hazard in relation to explosion likelihood.

5.2 EXAMPLE: LIKELIHOOD OF OCCURRENCE AND PREVENTION

Consider the issue of ignition source prevention. While every effort should, of course, be made to eliminate potential ignition sources in the workplace, it is widely acknowledged that the abundance and availability of industrial ignition sources render such efforts ineffective as the sole means of explosion prevention for gases. This explains, in part, the need for explosion-proof electrical equipment when dealing with flammable gases (i.e., the molecules of fuel and oxidant cannot effectively be separated from electrical ignition points). With combustible dusts on the other hand, it may be possible to effectively isolate

TABLE 5.1 Important Dust Explosibility Parameters [5]

Parameter	Typical units[a]	Description[b]	Example industrial applications[c]
P_{max}	bar(g)	Maximum explosion pressure in a constant-volume explosion	Containment Venting Suppression Isolation Partial inerting
$(dP/dt)_{max}$	bar/s	Maximum rate of pressure rise in a constant-volume explosion	As per P_{max}
K_{St}	bar·m/s	Size- or volume-normalized (standardized) maximum rate of pressure rise in a constant-volume explosion	As per P_{max}
MEC	g/m³	Minimum explosible (or explosive) dust concentration	Control of dust concentrations
MIE	mJ	Minimum ignition energy of a dust cloud (electric spark)	Removal of ignition sources Grounding and bonding
MIT	°C	Minimum ignition temperature of a dust cloud	Control of process and surface temperatures (dust clouds)
LIT	°C	Minimum ignition temperature of a dust layer or dust deposit	Control of process and surface temperatures (dust layers)
LOC	volume %	Limiting oxygen concentration in the atmosphere for flame propagation in a dust cloud	Inerting (with inert gas)

[a]The units given for P_{max} indicate a gauge (g) pressure or overpressure. One may also see P_{max} reported in units of absolute (a) pressure. Other pressure units such as pounds per square inch or psi (lb_f/in^2) are occasionally used, as are concentration units of ounces per cubic foot (oz/ft^3), and temperature units of °F.
[b]These parameters are described more fully in Chapters 18 and 19.
[c]See also Chapter 15.

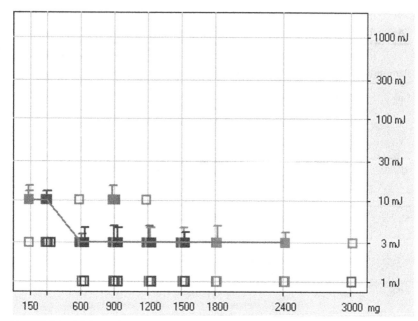

FIGURE 5.2 Minimum ignition energy (MIE) data for niacin (nicotinic acid) as determined in a MIKE 3 apparatus.

particulate material from ignition sources presented by electrical equipment. Differences in ignition likelihood for gases and dusts are typically reflected in the various hazardous area classification schemes adopted throughout the world.

As a specific example, Figure 5.2 gives a screenshot of minimum ignition energy (MIE) data for niacin (nicotinic acid) as determined in our laboratory during a recent round-robin apparatus calibration exercise. Dust amount (dispersed into a 1.2-L volume) appears on the x-axis and spark energy on the y-axis; the open boxes indicate no ignition at the particular ignition delay time and the solid boxes indicate ignition. What looks like an error bar on top of some of the boxes is in fact an indication of the number of attempts required to achieve ignition; ten repeat tests at a given set of conditions were performed before non-ignition was concluded. (See also Chapter 8.)

The data in Figure 5.2 demonstrate that the MIE of this sample is > 1 mJ and ≤ 3 mJ. Many flammable gases have ignition energies significantly less than 1 mJ. This is hardly the point, though, if one is handling niacin in an environment where electrostatic sparks having energies greater than the MIE abound. What is important is the dust explosion risk presented by the processing conditions in relation to the dust explosion hazard presented by the material itself.

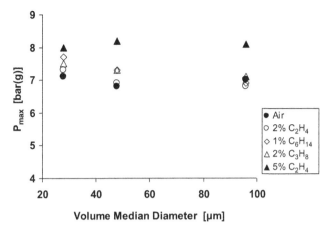

FIGURE 5.3   Maximum explosion pressure data for polyethylene and polyethylene/hydrocarbon gas hybrid mixture explosions in a Siwek 20-L chamber [6].

5.3 EXAMPLE: SEVERITY OF CONSEQUENCES AND MITIGATION

Figure 5.3 illustrates the results of laboratory-scale testing conducted by Amyotte et al. [6]. One can see a distinct range of P_{max} values between approximately 7 bar(g) and 8 bar(g) for the various experimental conditions. The lower boundary is for polyethylene dust alone at three different particle sizes. The upper boundary is essentially that of a stoichiometric or near-stoichiometric ethylene/air mixture (as would be attained by polyethylene in the presence of 5 volume % ethylene), and is therefore essentially representative of a gas-phase explosion. The data between these overpressure boundaries correspond to hybrid systems of dust with admixed ethylene, hexane, and propane (each gas being present in a concentration less than its respective lower flammable limit). The topic of hybrid mixtures is further developed in Section 5.4.

In all cases shown in Figure 5.3, the expected overpressure for an explosion in an enclosed volume is much greater than that able to be withstood by typical industrial-scale process equipment (which might be as much as two orders of magnitude lower). This knowledge, coupled with corresponding rate of pressure rise data, can be used to develop adequate explosion protection (mitigation) measures such as the installation of pressure relief panels. Again, with dust-handling equipment the relevant risk considerations would involve the relationship among explosibility parameters, vessel strength, and processing conditions.

5.4 HYBRID MIXTURES

As introduced in the preceding section, hybrid mixtures consist of a flammable gas and a combustible dust, each of which may be present in an amount less than the lower flammable limit (LFL) and minimum explosible concentration (MEC), respectively, and still give rise to an explosible mixture. The focus when discussing hybrid mixtures is often on admixture of a flammable gas in concentrations below the lower flammable limit of the gas itself, to an already explosible concentration of dust [4].

The influence of the co-presence of a flammable gas on the explosibility parameters of a fuel dust alone is well established [2]. These effects include higher values of maximum explosion pressure and maximum rate of pressure rise, and lower values of minimum explosible concentration and minimum ignition energy. There is, of course, already a hazard that exists when an explosible dust is present in a quantity above its minimum explosible concentration. With flammable gas admixture, the scenario is now one of magnification of an already-existing hazard, not the creation of a problem that did not exist in some form already.

Perhaps the most well-known hybrid mixture is the methane/coal dust system often encountered in underground coal mining. There are also several examples of hybrid mixture formation in other industries, such as the natural gas/fly ash system in fossil-fuel burning power plants and various hydrocarbon/resin combinations occurring in the production of plastic powders.

Another industry susceptible to hybrid mixture explosions is the production of pharmaceuticals, which often involves the transfer of combustible powders to vessels already containing a flammable solvent [7]. In keeping with the current theme of assessing the risk of a given fuel type rather than making strict hazard comparisons between fuel types, the following passage from a recent paper [7] (pp. 1–2) by Martin Glor is presented (with underlining added for emphasis):

… the probability of an explosion occurring during the transfer process is high because the probability of a coincidence in space and time of an explosive atmosphere and the activation of an effective ignition source, such as static electricity, is high. Furthermore, the severity of such an explosion could be disastrous, especially when taken into consideration the number of operators that would be directly exposed to the initial blast wave and subsequent fireball. Serious if not life threatening burns are likely, especially in the presence of a dust cloud or hybrid mixture explosion.

5.5 REALITY

Unintentional explosions of flammable gases and combustible dusts respectively represent undesirable occurrences. Whether one fuel type is by nature "worse" than the other is somewhat of a moot point. The real issue is the risk posed by a given fuel hazard (gaseous or particulate) expressed in terms of explosion likelihood and consequence severity for a given scenario. Only with these considerations in mind can appropriate prevention and mitigation measures be designed and implemented.

As a final example, Figure 5.4 gives explosion pressure data for a gaseous fuel (methane) and two dusts (polyethylene and coal). The pressure data for the

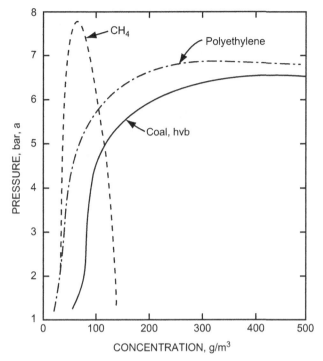

FIGURE 5.4   Explosion pressure data for methane compared with polyethylene and coal (high-volatile bituminous or hvb) dusts [2].

gas and dusts are comparable to those given in Figure 5.3, so the discussion in Section 5.3 would also hold here. The additional information imparted by Figure 5.4 is the absence of an upper (rich) flammable limit for the dusts and the existence of such a limit for methane—corresponding to its established upper flammable limit of 15 volume % expressed as a mass concentration in units of g/m^3 [2].

What this means practically is that activities such as dense-phase pneumatic conveying of powders are carried out using dust concentrations higher than an upper flammable limit that is near impossible to measure. This does not mean that in this regard dust explosions are "worse" than gas explosions; it simply means that an additional factor has been introduced to the assessment of dust explosion risk.

5.6 WHAT DO *YOU* THINK?

In this chapter we have talked at length about *risk* and how it is composed of *probability* (or *likelihood*) of occurrence and *severity* of consequences. As scientists and engineers, we sometimes forget to our great disadvantage that there is a third, often hidden, component of risk—*perception*. How many

FIGURE 5.5 Newspaper photograph of Halifax forest fire on May 12, 2012, showing a residential area, grain elevator, and cruise ship [8]. *(Reprinted with permission from the Halifax Herald Limited.)*

industrial enterprises have been announced with proclamations of acceptable risk levels only to be delayed by stakeholder concern or derailed by public outrage?

Consider the picture in Figure 5.5, which shows a forest fire that occurred in my hometown of Halifax, Nova Scotia, on May 12, 2012. The distances are a bit deceiving, but in the vicinity of the fire, there were several residences; located on the harborfront are a grain elevator and a port housing a cruise ship facility. You would not likely be surprised to learn that the focus of reported emergency response was on the affected residents and their dwellings. However, I recall nothing in the media reports at the time concerning potential threats to the grain elevator or cruise ship. Then again, there are people living close by the grain-handling facility year-round and cruise ships in the vicinity on a seasonal basis. It also happens there was an explosion at the grain elevator in 2003 that caused significant on-site property damage and necessitated the evacuation of hundreds of people from their homes.

All this makes me wonder about risk and the perception of what is acceptable. How about you?

Although I would argue that the earlier exercise is indeed quite technical, perhaps you would like to try something a bit more "technically" in line with this chapter's discussion on dust and hybrid mixture explosions. Recent papers [9,10] from researchers in Italy have brought new insight to the issue of gas, dust, and hybrid mixture explosion severity as expressed by the size-normalized maximum rate of pressure rise, or K_{St}, parameter (see Table 5.1). I would invite interested readers to source these journal articles and explore the matter further.

As an example of this work, here we look at Figure 5.6, which shows an explosion regime chart for niacin dust and acetone gas [10]. (The paper by Garcia-Agreda et al. [9] gives a similar chart for niacin and methane.) Full details are given by Sanchirico et al. [10]; briefly, the x-axis in Figure 5.6 shows

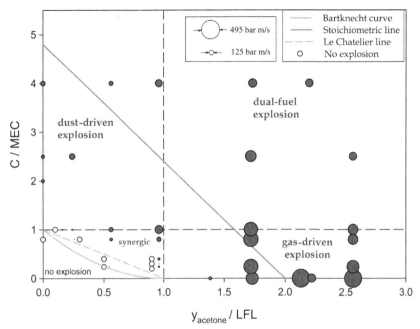

FIGURE 5.6 Explosion regimes for mixtures of niacin dust and acetone gas as determined in a Siwek 20-L chamber [10].

the ratio of acetone concentration to its lower flammable limit while the y-axis shows similar data for the niacin dust.

Five regimes are displayed by Figure 5.6: (i) lower part of lower-left corner—region where both the gas and dust concentrations are less than the respective LFL and MEC, and too low to support combustion; (ii) upper part of lower-left corner—region where both the gas and dust concentrations are less than the respective LFL and MEC, yet high enough to lead to an explosion, (iii) upper-left corner—explosion region where the gas concentration is less than the LFL and the dust concentration exceeds the MEC; (iv) upper-right corner—explosion region where both the gas and dust concentrations exceed the respective LFL and MEC; and (v) lower-right corner—explosion region where the gas concentration exceeds the LFL and the dust concentration is less than the MEC. Also indicated by the size of the data circles in Figure 5.6 is the magnitude of the K_{St} values.

At first glance, Figure 5.6 might seem to support the concept of gas explosions being nominally worse than dust explosions. After all, the data demonstrate that gas-driven explosion severities (K_{St} values) for this fuel system are greater than those measured for synergic, dust-driven, and dual-fuel explosions. Consistent with the discussion in this chapter, however, such an observation is limited to a discussion of the hazard presented by the various flammable gas/combustible dust combinations. If an explosion were to occur in an industrial setting according to any regime shown in Figure 5.6, the rates of pressure rise indicated—if not mitigated—would likely be high enough to cause harm to people and damage to property. With these considerations in mind, it becomes clear that process risk (not just material hazard) must be a dominant concern.

REFERENCES

[1] Eckhoff RK. Dust explosions in the process industries, 3rd ed. Boston, MA: Gulf Professional Publishing/Elsevier; 2003.

[2] Cashdollar KL. Overview of dust explosibility characteristics. Journal of Loss Prevention in the Process Industries 2000;13:183–99.

[3] Wilson L, McCutcheon D. Industrial safety and risk management. Edmonton, AB, Canada: University of Alberta Press; 2003.

[4] Amyotte PR, Eckhoff RK. Dust explosion causation, prevention and mitigation: an overview. Journal of Chemical Health & Safety 2010;17:15–28.

[5] Amyotte PR. Dust explosions happen because we believe in unicorns. Keynote Lecture, Proceedings of 13th Annual Symposium, Mary Kay O'Connor Process Safety Center, Texas A&M University, College Station, TX; October 26–28, 2010. pp. 3–30.

[6] Amyotte P, Lindsay M, Domaratzki R, Marchand N, Di Benedetto A, Russo P. Prevention and mitigation of dust and hybrid mixture explosions. Process Safety Progress 2010;29:17–21.

[7] Glor M. A synopsis of explosion hazards during the transfer of powders into flammable solvents and explosion preventative measures. Pharmaceutical Engineering 2010;30:1–8.

[8] Chronicle Herald. http://thechronicleherald.ca/metro/99017-purcells-cove-fire-no-longer-threat-after-rain-officials-say; 2012 (last accessed September 9, 2012).

[9] Garcia-Agreda A, Di Benedetto A, Russo P, Salzano E, Sanchirico R. Dust/gas mixtures explosion regimes. Powder Technology 2011;205:81–6.

[10] Sanchirico R, Di Benedetto A, Garcia-Agreda A, Russo P. Study of the severity of hybrid mixture explosions and comparison to pure dust-air and vapour-air explosions. Journal of Loss Prevention in the Process Industries 2011;24:648–55.

Chapter 6

Myth No. 5 (Fuel): It's Up to the Testing Lab to Specify Which Particle Size to Test

Of all possible material characteristics, particle size has arguably the greatest impact on the explosibility parameters described in Chapter 5. Other fuel properties are undeniably important—e.g., particle shape and moisture content—but it is particle size that largely determines sample reactivity and therefore explosibility.

In this chapter we first examine why particle size is such an important concern in the prevention and mitigation of dust explosions. Several examples are given to illustrate the role of particle size so as to establish a common level of understanding. The discussion then shifts to who should specify sample particle sizes for testing purposes (i.e., determination of parameters like P_{max} and K_{St}). Is it the off-site testing lab personnel? External consultants? Company management? Process operators? Loss prevention engineers on-site?

My view on this is quite simple. It is not the testing lab; at least not in isolation from those for whom the testing is being done. It is highly unlikely and certainly not advisable that specification of the locations for flammable gas detectors would be solely entrusted to someone who has never visited the facility being protected and has no knowledge of the processing conditions involved. Why then would specification of dust particle size be handed over entirely to someone having similar qualifications?

Figure 6.1 shows two wood samples such as might be encountered in an industrial application. Should the sample shown in the lower photograph be ignored because its size is larger than the sample appearing in the upper photograph? Who should answer this question? And are smaller sizes of wood dust present in the facility (say, −200 mesh or < 75 μm)? If so, should this size be tested for explosibility parameters?

Similarly, Figure 6.2 shows samples of polyethylene both as it emerges from the reactor and after pelletizing for shipment to secondary manufacturers. A conclusion might reasonably be reached not to test the pellets given that the purpose of this formulation is, in part, to address the issue of dust explosions during transport. But what about the product exiting the reactor, or finer material that might be carried overhead in the dust collection system? Should samples from these areas be collected and tested for explosibility? Who should make this decision?

6.1 ROLE OF PARTICLE SIZE DISTRIBUTION

Although it is common to use the term *dust particle size*, we are usually concerned more with the actual particle size *distribution*. Few dusts are mono-size (i.e., composed of particles having a single diameter or length-to-diameter ratio), or have size distributions so well characterized that the dust can be completely specified simply by name.

Exceptions do occur. If an industrial process is designed to produce a narrow-size powder, then of course, the end-product will have a well-defined and

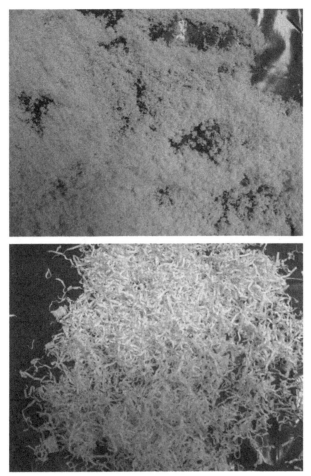

FIGURE 6.1   Upper: digital photograph of −35 mesh (< 500 μm) wood sample; Lower: digital photograph of +35 mesh (> 500 μm) wood sample [1].

narrow distribution. Wider size distributions might exist, however, if at some stage in the process pulverizing and sieving of a bulk material are involved.

Lycopodium, a naturally occurring spore of the club moss [2], has an almost uniform particle diameter of 30 μm [3]. Variation of particle size is thus eliminated in experimental programs in which lycopodium is used as the test dust. Pittsburgh seam bituminous coal dust is another material often used as a reference dust for test purposes and has a size distribution of ~80% −200 mesh (< 75 μm) and ~50% −325 mesh (< 45 μm), with a volatile content of 36% [4].

In most cases, dust size must be characterized by giving its distribution in volume or weight (mass) percentages or mesh sizes; with mesh size distributions, *plus* (+) means that the dust does not pass through the particular screen,

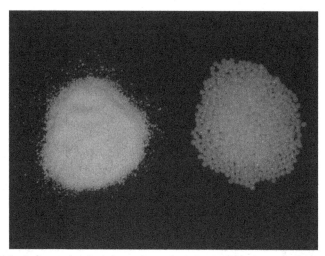

FIGURE 6.2   Left: sample of reactor-product polyethylene (volume mean diameter = 900 μm);
Right: sample of polyethylene in pellet form for transportation off-site.

whereas *minus* (–) indicates the converse. Sometimes only the mean or median
particle diameter is reported on a mass or volumetric basis; some indication of
the width of the actual size distribution still needs to be given in these cases.
(Note that mass and volumetric bases are the same with an assumption of constant density throughout the entire size distribution.)

It is sometimes said that narrow size distribution dusts are more fundamental, whereas wide size distribution dusts are more practical. What this means
is that identification of intrinsic features in laboratory testing may be possible
only with narrow size dusts, whereas information on the explosibility of wide
size range dusts may be more relevant to industrial applications. For example,
Hertzberg et al. [5] determined that the minimum explosible concentration of
the aforementioned Pittsburgh seam bituminous coal dust is independent of particle size for mass mean diameters less than approximately 50 μm. It is essential
to note, however, that their study was performed with coal dust samples having
narrow size distributions [5]. The appearance of such a *characteristic diameter*
would not likely occur with much wider size distributions of the same material.
This point again reinforces the necessity of acquiring test data for the actual dust
encountered in an industrial process and not relying solely on literature values
of explosibility parameters for the generic material. (See Section 2.2.)

Why then is particle size such an important factor in determining the dust
explosion hazard of a given material? Simply put, a decrease in particle size
leads to an increase in particle surface area and therefore an enhancement of
dust reactivity. If combustion is between gaseous oxidant and solid fuel (i.e.,
heterogeneous combustion), then more active sites for chemical reaction are
available on the surface of each particle. If combustion is between gaseous

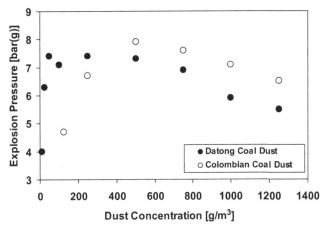

FIGURE 6.3   Explosion pressure as a function of dust concentration for coal dust as determined in a Siwek 20-L chamber [6].

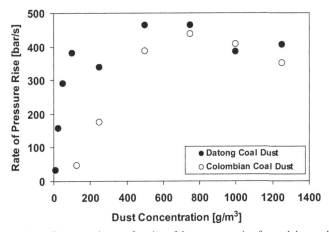

FIGURE 6.4   Rate of pressure rise as a function of dust concentration for coal dust as determined in a Siwek 20-L chamber [6].

oxidant and gaseous fuel that has evolved by melting/vaporization or pyrolysis/ devolatilization of the particulate matter (i.e., homogeneous combustion), then the rates of these gaseous fuel evolution processes are enhanced by the increase in particle surface area. An example will illustrate these points.

As noted previously (e.g., Chapter 3), coal mining inevitably leads to the generation of coal dust deposits in working galleries. The particle size of the coal dust has a profound influence on its explosibility as illustrated in Figures 6.3 and 6.4, which provide data for float coal dust from the Datong mine in China. These tests were conducted in a laboratory-scale chamber using a 5-kJ ignition energy. The coal dust had the physical properties shown in Table 6.1.

TABLE 6.1 Proximate Analysis and Particle Size Distribution of Datong Float Coal Dust [6]

Moisture [weight %]	Ash [weight %]	Volatiles [weight %]	Fixed carbon [weight %]	Particle size distribution [weight %]
0.1	11.6	25.8	62.5	100% < 75 μm; 98% < 45 μm; 78% < 20 μm; 51% < 10 μm

TABLE 6.2 Proximate Analysis and Particle Size Distribution of Colombian Coal Dust [7]

Moisture [weight %]	Ash [weight %]	Volatiles [weight %]	Fixed carbon [weight %]	Particle size distribution [weight %]
0.4	7.7	36.4	55.5	83% < 125 μm; 73% < 75 μm; 53% < 45 μm; 29% < 20 μm

As seen in Figures 6.3 and 6.4, the Datong coal dust explosibility data are comparable to those from similar laboratory-scale (5-kJ) testing for a larger-particle-size, higher-volatile-content Colombian coal (Table 6.2), except at the lower end of the dust concentration range. The explosion hazard is therefore significantly magnified by the presence of very fine coal dust in a mine. The data in Figures 6.3 and 6.4 are explained by the exceedingly small size of the Datong dust, leading to rapid devolatilization of the particles and generation of combustible volatiles at such a high rate that the actual combustion of the volatiles becomes the rate-limiting step in the overall explosion process.

6.2 PARTICLE SIZE EFFECTS ON EXPLOSIBILITY PARAMETERS

Amyotte et al. [6] provide numerous examples of the influence of particle size on the dust explosibility parameters shown previously in Table 5.1. The emphasis in their work is on the possible safety benefits of a deliberate increase in dust particle size as would be brought about by the inherent safety principle of *moderation* (discussed at length in Chapter 15).

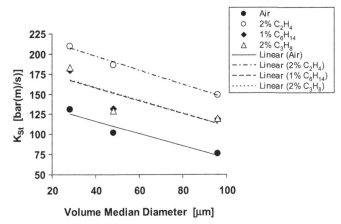

FIGURE 6.5   Size-normalized maximum rate of pressure rise data for polyethylene and polyethylene/ hydrocarbon gas hybrid mixture explosions in a Siwek 20-L chamber [11].

Here, our focus is on the enhanced reactivity that accompanies a decrease in dust particle size—the effects of which are numerous and well established. They include, for example, an increase in the maximum explosion pressure, P_{max}; a potentially significant increase in the maximum rate of pressure rise, $(dP/dt)_{max}$ (and hence, K_{St}); and a decrease in the minimum explosible concentration (MEC), minimum ignition energy (MIE), and minimum ignition temperature (MIT) [8,9]. Decreasing the mean diameter of an explosible dust by decreasing the particle size throughout the entire size distribution can also significantly increase the inerting level [10]. (See Section 4.4 and Chapter 7.)

Figure 6.5 gives size-normalized maximum rate of pressure rise (K_{St}) data for the same combinations of high-density polyethylene and hydrocarbon gas as shown in Figure 5.3 (with the exception of 5 volume % ethylene). One can clearly see that the step-wise combustion process for polyethylene powder is rendered more efficient by a reduction in particle size or admixture of flammable gas, both of which act to partially negate the potentially rate-limiting step of volatiles generation by particle heating, melting, and vaporization.

K_{St}, as introduced earlier in this book, is the size-normalized maximum rate of pressure rise for a constant-volume explosion, as determined in standardized equipment according to standardized test procedures. It is the product of the maximum rate of pressure rise, $(dP/dt)_{max}$, and the cube-root of the volume, V, of the chamber in which the explosion tests were conducted. The units of K_{St} are therefore units of (pressure)(length)/(time); conventional units are bar·m/s as indicated in Table 5.1. K_{St} can also be thought of as being numerically equivalent to the maximum rate of pressure rise occurring in a standardized 1-m³ test chamber for a given dust sample. This parameter will appear many times in the remainder of the book and is discussed extensively in Chapters 18 and 19.

A recent study by Nifuku et al. [12] was directed toward an investigation of particle size influence on MEC, MIE, and MIT for aluminum and magnesium dusts generated during shredding processes involved in industrial recycling. As illustrated in Table 6.3 for magnesium, both MEC and MIE were observed to decrease exponentially with a decrease in particle size for the narrow size distributions used; MIT was also found to be directly dependent on particle size.

Data such as those given in Figures 5.3 and 6.5 are used for a number of purposes, one of which is the sizing of explosion relief vents (a passive-engineered safety measure as discussed in Chapter 15). Selection of an appropriate particle size (or more properly, particle size distribution) for testing is therefore critically important in the design of explosion protection measures. This point is illustrated by Table 6.4, which shows explosibility data for dust generated in a wood processing facility. In the actual process, although the coarse dust was predominant, pockets of the fine dust were found in a dead-space in a process unit header [6]. It is clear that a vent design based on the K_{St} value of the coarse dust would be inadequate for protection from the effects of a dust explosion involving the fine wood dust.

Thus, the selection of a dust sample for testing to see if the material represents an explosion hazard should be preceded by considerations related to process risk. For example, in plants with dust extraction systems, it is typical to select dust samples from filters. These samples will be finer than the main product and, hence, will explode more violently and ignite more readily. Here,

TABLE 6.3 Magnesium Dust Explosibility Results [12]

Size fraction [μm]	MEC [g/m³]	MIE [mJ]	MIT [°C]
149–177	900	240	620
1–20	90	4	520

TABLE 6.4 Effect of Particle Size on K_{St} Values of Wood Dust Samples as Determined in a Siwek 20-L Chamber [6]

Dust	Particle size distribution [weight %]	K_{St} [bar·m/s]
Coarse	50% < 1 mm 0.3% < 75 μm	9
Fine	93% < 1 mm 35% < 125 μm 16% < 75 μm	130

the reasoning behind the selection of the sample to be tested is directly or indirectly a part of the assessment of the process risk [13]. These considerations help to explain the rationale behind testing standards such as ASTM E1226 [4], which states that although tests may be conducted on an as-received sample, it is recommended that the test sample be at least 95% –200 mesh (< 75 μm). The reason given is the possible accumulation of fines at some location in a processing system [4]; this is precisely the situation described previously for the data in Table 6.4.

6.3 A COOPERATIVE ENDEAVOR

ASTM E1226 [4] and other national and international standards can only provide guidance on dust samples for testing. They cannot be expected to address every possible nuance of every industrial facility. Similarly, it should not be left entirely to the explosion testing facility to specify the appropriate particle size distribution for test purposes.

Some test labs are just that—facilities housing explosion testing equipment and which perform testing services on samples received. Others offer external consultancy services that may extend to providing advice on locations for sample collection (often in conjunction with a site visit). Sometimes the capabilities are distributed and an industrial plant will be dealing with both a test facility and a separate external consultant.

Regardless of the arrangement, the selection of sample collection points and size distribution(s) to be tested must be a cooperative endeavor among the various parties. An industrial colleague recently remarked to me that having worked extensively with contract dust testing labs, she has a good appreciation of the need for the client to work in partnership with these facilities on the matter of sample characterization (Prine, B., 2012. Personal communication, with permission).

Table 6.5 gives the results of one such partnership. In the course of designing a new process plant (i.e., one for which there was no prior operating experience),

TABLE 6.5 Effect of Particle Size on Explosibility Parameters of Wood Dust Samples as Determined in a Siwek 20-L Chamber (P_{max} and K_{St}) and a MIKE 3 Apparatus (MIE)

Dust	Particle size distribution [weight basis]	P_{max} [bar(g)]	K_{St} [bar·m/s]	MIE [mJ]
Sample A (coarse)	75 μm–500 μm	7.7	51	> 1000
Sample A (fine)	< 75 μm	8.7	188	10–30
Sample B (coarse)	75 μm–500 μm	8.1	63	> 1000
Sample B (fine)	< 75 μm	8.8	200	10–30

the designers, loss prevention consultants, and testing lab personnel collectively decided to adopt a differentiated approach for determination of dust explosion parameters. This involved testing coarse and fine samples that were well defined in terms of particle size distribution and represented the extremes of expected processing conditions. The data in Table 6.5 also formed the basis for testing considerations related to other potential size distributions and the preventive and protective measures required.

6.4 REALITY

As discussed previously in Chapter 5, there is a need to distinguish between material *hazard* and process *risk* when dealing with dust explosion phenomena. Because particle size distribution is one of the key properties of a dust that defines its material hazard, knowledge of whether a particular size distribution is actually encountered in a given application is required to assess the process risk. This highlights the importance of gaining a thorough understanding of the dust handling process under consideration during both normal and upset conditions. A dust explosion test facility or consultancy can help with general considerations in this regard, but specific input on the selection of samples for testing must come from experienced plant personnel.

Although it may be convenient to identify a particle size distribution by a single value of the mean or median diameter, information on the entire size range must also be readily available. An innovative experimental study in this regard was recently carried out by Castellanos et al. [14] at the Mary Kay O'Connor Process Safety Center, Texas A&M University. These authors introduced the term *polydispersity* to the dust explosion research and practice communities.

Polydispersity is used in the powder characterization literature to indicate the degree of heterogeneity of particle size, and is defined by Castellanos et al. [14] as

$$\sigma_D = (D_{90} - D_{10})/D_{50} \tag{6.1}$$

where σ_D is the particle size polydispersity, D_{90} is the diameter greater than that of 90 volume % of the particles, D_{10} is the diameter greater than that of 10 volume % of the particles, and D_{50} is the diameter greater than that of 50 volume % of the particles (i.e., the volume median diameter).

Experimental results were obtained for five aluminum dust samples having essentially the same D_{50} (14.2–14.9 μm) and σ_D values ranging from 0.95–2.51. The data clearly showed the effects of polydispersity on explosibility; the sample with the lowest σ_D had a significantly lower value of K_{St} and higher value of MIE than the sample with the highest σ_D. The important conclusion from this work is—in the authors' own words—"that simply stating the mean diameter of a dust inadequately describes its hazard potential [14]."

6.5 WHAT *YOU* THINK?

In keeping with the concept of relating the text material to practice in your own facility, you might consider arranging for dust to be sampled from different locations in your plant. Are there differences in the size distributions of the samples collected from these locations? What impact might these size differences be expected to have on explosion likelihood parameters such as MEC, MIE, and MIT? Or on consequence severity parameters such as P_{max} and K_{St}?

If you already have an established dust sampling program, have you reviewed it recently to ensure process changes have not introduced different-sized material to locations previously not monitored? Has there been a trend to finer sizes of dust being produced at your facility? If so, have any necessary modifications been made to existing prevention and mitigation measures?

As a final exercise, a brief introduction to dust explosion modeling is presented. The intention is not to delve too deeply into modeling of dust explosion phenomena, but rather to present a few model results and relate them to our discussion of particle size effects.

Figure 6.6 gives K_{St} values for polyethylene as determined by laboratory-scale experimentation and the thermo-kinetic model of Di Benedetto et al. [15] Do the model results for K_{St} follow the expected trend with a change in particle diameter?

Figure 6.7 shows polyethylene dust explosion simulations performed using the computational fluid dynamics (CFD) software Dust Explosion Simulation Code (DESC) developed by GexCon AS in Norway [16]. Note that the pressure/time trace on the left is for a smaller particle size distribution than the trace on the right in Figure 6.7. Does the change in overpressure (y-axis) match your

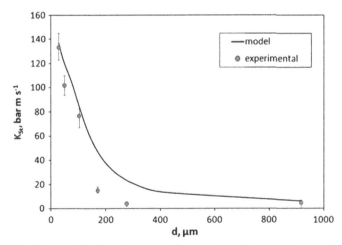

FIGURE 6.6 Size-normalized maximum rate of pressure rise for polyethylene as a function of dust diameter as obtained by experimentation and thermo-kinetic modeling [15].

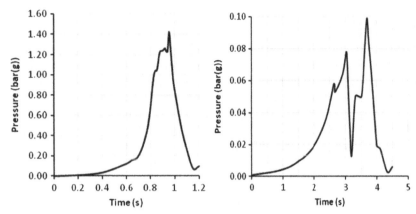

FIGURE 6.7　DESC simulations for polyethylene dust explosions in a 400-m³ storage silo with a dust cloud concentration of 500 g/m³. Left: –200 mesh (< 75 μm); Right: –70 mesh (< 212 μm) [17].

expectation given the change in particle size? What about the change in time (x-axis) to reach the peak overpressure?

These may seem like trivial questions, but I would argue that they are quite critical. If a model cannot match the trends in the most basic of dust explosion phenomena, then what hope is there that the actual model-predicted values will bear any relation to reality?

REFERENCES

[1] Amyotte PR, Cloney CT, Khan FI, Ripley RC. Dust explosion risk moderation for flocculent dusts. Journal of Loss Prevention in the Process Industries 2012;25:862–9.

[2] Landau SI, editor. Funk and Wagnall's standard college dictionary. New York, NY: Funk and Wagnalls; 1968.

[3] Mason WE, Wilson MJG. Laminar flames of lycopodium dust in air. Combustion and Flame 1967;11:195–200.

[4] ASTM. ASTM E1226–10, Standard test method for explosibility of dust clouds. West Conshohocken, PA: American Society for Testing and Materials; 2010.

[5] Hertzberg M, Cashdollar KL, Zlochwer I, Ng DL. Domains of flammability and thermal ignitability for pulverized coals and other dusts: particle size dependences and microscopic residue analyses. 19th Symposium (International) on Combustion, Pittsburgh, PA: The Combustion Institute; 1982. pp. 1169–80.

[6] Amyotte PR, Pegg MJ, Khan FI, Nifuku M, Yingxin T. Moderation of dust explosions. Journal of Loss Prevention in the Process Industries 2007;20:675–87.

[7] Amyotte PR, Basu A, Khan FI. Reduction of dust explosion hazard by fuel substitution in power plants. Process Safety and Environmental Protection 2003;81:457–62.

[8] Cashdollar KL. Overview of dust explosibility characteristics. Journal of Loss Prevention in the Process Industries 2000;13:183–99.

[9] Amyotte PR, Mintz KJ, Pegg MJ, Sun YH, Wilkie KI. Laboratory investigation of the dust explosibility characteristics of three Nova Scotia coals. Journal of Loss Prevention in the Process Industries 1991;4:102–9.

[10] Amyotte PR, Mintz KJ, Pegg MJ, Sun YH, Wilkie KI. Effects of methane admixture, particle size and volatile content on the dolomite inerting requirements of coal dust. Journal of Hazardous Materials 1991;27:187–203.

[11] Amyotte P, Lindsay M, Domaratzki R, Marchand N, Di Benedetto A, Russo P. Prevention and mitigation of dust and hybrid mixture explosions. Process Safety Progress 2010;29:17–21.

[12] Nifuku M, Koyanaka S, Ohya H, Barre C, Hatori M, Fujiwara S, Horigichi S, Sochet I. Ignitability characteristics of aluminum and magnesium dusts relating to the shredding processes of industrial wastes. Halifax, Canada: Proceedings of Sixth International Symposium on Hazards, Prevention, and Mitigation of Industrial Explosions; August 27–September 1, 2006.

[13] Amyotte PR, Eckhoff RK. Dust explosion causation, prevention and mitigation: an overview. Journal of Chemical Health & Safety 2010;17:15–28.

[14] Castellanos D, Carreto V, Mashuga C, Trottier R, Mannan SM. The effect of particle size dispersity on the explosibility characteristics of aluminum dust. Krakow, Poland: Proceedings of Ninth International Symposium on Hazards, Prevention, and Mitigation of Industrial Explosions; July 22–27, 2012.

[15] Di Benedetto A, Russo P, Amyotte PR, Marchand N. Modelling the effect of particle size on dust explosions. Chemical Engineering Science 2010;65:772–9.

[16] Skjold T. Review of the DESC project. Journal of Loss Prevention in the Process Industries 2007;20:291–302.

[17] Abuswer M, Amyotte P, Khan F. A quantitative risk management framework for dust and hybrid mixture explosions. Journal of Loss Prevention in the Process Industries 2013;26: 283–9.

Chapter 7

Myth No. 6 (Fuel/Ignition Source): Any Amount of Suppressant Is Better Than None

Inert dusts (or suppressants) are used in industry in both prevention and mitigation roles; the distinction in their use—whether it is to address explosion likelihood or explosion consequences—is explained in Section 7.1. Regardless of the application, it may be tempting to conjecture that even a small amount of inert dust will have a beneficial impact. After all, the added dust is noncombustible, and one might therefore think it reasonable to conclude that risk has been reduced to an acceptable level by its admixture with the combustible dust. Regrettably, this is not the case.

In the current chapter, we turn our attention for the first time in this book to an explosion pentagon element other than the *fuel* itself. Fuel considerations still are involved in our discussion here, but *ignition source* and the related issue of flame propagation temperature now enter the picture more fully. As shown in the subsequent sections (after the means and mechanisms of dust flame extinction are first examined), use of an inert dust in amounts less than that required for complete flame extinction can provide a false sense of security and can actually raise the level of consequence severity.

Figure 7.1 presents an introductory example in this regard. Equating the number of suppressant containers with the quantity of applied suppressant, one clearly sees that only a sufficient amount of extinguishing powder injected in a timely manner can achieve the desired result of overpressure limitation via explosion suppression.

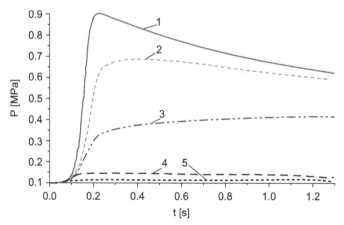

FIGURE 7.1   Pressure/time traces for unsuppressed, partially suppressed, and suppressed explosions of cornstarch (260 g/m^3) in a 1.25-m^3 explosion chamber [1]. Legend for traces: 1—without suppression; 2—one suppressant container (triggering pressure = 1.1 bar); 3—one suppressant container (triggering pressure = 1.015 bar); 4—two suppressant containers (triggering pressure = 1.015 bar); 5—two suppressant containers (triggering by photodiode signal).

7.1 INERTING AND SUPPRESSION

The term *inerting* arises when focusing on *preventing* the occurrence of a dust explosion; use of an inert dust admixed in sufficient amount with a fuel dust can lead to removal of the heat necessary for combustion [2]. This is somewhat analogous to explosion prevention by use of an inert gas (e.g., nitrogen) to ensure that process operation occurs at oxygen concentrations below the maximum permissible level. (See Chapter 10.)

On the other hand, the term *suppression* arises when focusing on *mitigating* the consequences of a dust explosion [2]. The intent is the same as with inerting—to remove the heat necessary for sustained combustion and thus to limit the generation of destructive overpressures in an enclosed volume. In the case of suppression, however, the inert dust is injected into the just-ignited explosible dust/air mixture rather than being intimately premixed with the explosible dust prior to ignition, as in the case of inerting.

Inerting or suppression thus typically occurs via a thermal quenching mechanism in which the temperature in the region of incipient combustion is reduced to a value below the limiting flame temperature for self-sustained chemical reaction and flame propagation [3]. Some inert powders used for this purpose may also act to chemically interfere with the combustion process [3]. Hence, one often hears the terms *thermal* and *chemical* inhibitors.

The fundamental difference, as well as the linkage between inerting and suppression, is clearly shown in Figure 7.2 from Moore and Siwek [4]. As described

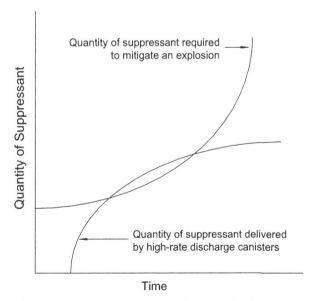

FIGURE 7.2 Suppressant (or inertant) requirement and delivery curves [4].

by Dastidar and Amyotte [5], Figure 7.2 is a pictorial representation of suppressant (or inertant) requirements to mitigate an explosion, coupled with suppressant delivery from high-rate discharge (HRD) canisters. The quantity of suppressant required to mitigate an explosion begins at some initial value at zero time and increases rapidly as the explosion fireball increases in size. This initial amount of suppressant is dependent on material properties and the nature of the ignition source. The upward curvature of the line is dependent on the nature of flame propagation through the explosible dust in a given vessel geometry.

The delivery of the suppressant from the discharge canisters begins after an initial delay that is dependent on the explosion detection system. The curve then increases rapidly and begins to level off as the suppressant propellant is spent. For adequate suppression of an explosion, the suppressant delivery curve must cross and exceed the suppressant requirement curve. The nature and curvature of the suppressant delivery curve are dependent on the design characteristics of the suppressant device (e.g., number and type of detectors, number and size of discharge canisters, and flow rate from the discharge canisters). A key factor in determining if the two curves cross is the intersection of the curve showing the quantity of suppressant required to mitigate an explosion, with the ordinate (i.e., the suppressant requirement at zero time). This value is the inerting level, or the suppressant requirement when the suppressant (or inertant) and the explosible dust are intimately mixed.

Figure 7.3 shows examples of dust explosion HRD systems having various suppressant volumes. Each extinguisher consists of a suppressant canister at the top separated by a membrane at the flanged section from the connecting tube feeding into the dispersing nozzle at the bottom. The plenary paper by Klemens [1] provides further details on such protective devices.

Whether the term *inertant* or *suppressant* is used, it is the same material employed—an inert (i.e., non-combustible) dust. The important difference is

FIGURE 7.3 High-rate discharge (HRD) suppressant systems with volumes of 2, 5, and 10 L (left to right in picture) [1].

in the application of the material—be it *inerting* or *suppression*. This same distinction is clearly articulated by Eckhoff [6] in his chapter on dust explosion research and development during the period 1990–2002, wherein he presents different sections titled "Inerting by Adding Noncombustible Dust" and "Automatic Explosion Suppression." An additional distinction elucidated in Chapter 15 of the current book is the identification of inerting (with an inert dust) as an inherent safety measure and suppression as an active engineered safety measure.

As noted by Eckhoff [6], suppression is more widely applicable than inerting in the process industries. This is due to the need to intimately premix the explosible and inert dusts when the application is explosion inerting; this mixing can lead to unacceptable product contamination. Inerting is perhaps most prevalent in the underground coal mining industry, where as previously described, inert rock or stone dust (e.g., limestone or dolomite) is mixed with coal dust generated during the mining process. Inert materials typically used for suppression purposes include sodium bicarbonate (abbreviated as SBC with chemical formula $NaHCO_3$) and monoammonium phosphate (abbreviated as MAP with chemical formula $NH_4H_2PO_4$).

7.2 MINIMUM INERTING CONCENTRATION

It was noted in the preceding section that some inertants have the ability to exert a chemical influence in addition to the mechanism of thermal energy absorption. These inhibitors chemically participate in the combustion sequence by acting to terminate chain branching reactions via free radical capture, thus providing kinetic interference in the process of flame propagation. Sodium bicarbonate (SBC) and monoammonium phosphate (MAP) are thought to act as explosion inhibitors by both thermal and chemical means [7]. Rock or stone dust (limestone, dolomite, etc.) is a thermal inhibitor.

Figure 7.4 shows the results of inerting experiments with limestone, SBC, and MAP as inertants and Pittsburgh pulverized coal as fuel. The primary purpose of these tests was to investigate a flammability parameter known as the minimum inerting concentration, or MIC, which is applicable to the case of inert dust admixed with a combustible dust prior to ignition.

The data points in Figure 7.4 represent the explosibility limit of the given fuel/inertant mixtures. They have been interpolated by averaging the highest inertant concentration tested that produced an explosion with the lowest inertant concentration tested that did not produce an explosion for a given fuel concentration. The area to the left of each curve in Figure 7.4 represents the explosible region for the fuel/inertant mixture. The area to the right of each curve represents the non-explosible region; here there is sufficient inertant to prevent an explosion. The "nose" of each curve (or envelope) represents the least amount of inertant that would prevent an explosion regardless of fuel concentration—i.e., the aforementioned MIC.

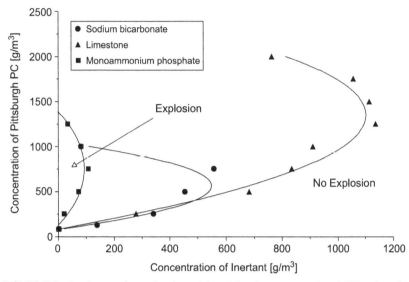

FIGURE 7.4 Inerting envelopes showing minimum inerting concentration (MIC) values for Pittsburgh pulverized coal dust with limestone, sodium bicarbonate, and monoammonium phosphate as inertants; experiments conducted in a spherical 1-m³ chamber [7,8].

Figure 7.4 clearly illustrates that for inerting of Pittsburgh pulverized coal, limestone is the least effective inhibitor and MAP is the most effective inhibitor. This occurrence is explainable by the inability of limestone to decompose in the rapidly advancing flamefront [7]. Additionally, the chemical inhibition properties of MAP and SBC would be expected to enhance their effectiveness over limestone. It should be noted that the superior performance of MAP over SBC illustrated in Figure 7.4 did not hold for all combinations of fuel and inertant tested [5,7–9]. This means that, as would be expected, an inertant's effectiveness is dependent on the composition of the explosible dust. Further, as demonstrated by Amrogowicz and Kordylewski [10], such effectiveness may also be dependent on whether the application is *inerting* or *suppression*. They found that for their suite of explosible dusts (melamine, wood dust, wheat flour, and coal dust), MAP was more effective than SBC for *inerting*, whereas SBC performed as well as or better than (depending on inhibitor concentration) MAP for *suppression*.

7.3 SUPPRESSANT ENHANCED EXPLOSION PARAMETER

The importance of ensuring that sufficient inertant is present during an explosion scenario is demonstrated by Figure 7.5 from the 1-m³ SBC inerting testing conducted by Dastidar et al. [8]. Here we observe a phenomenon termed SEEP—Suppressant Enhanced Explosion Parameter—where the explosion

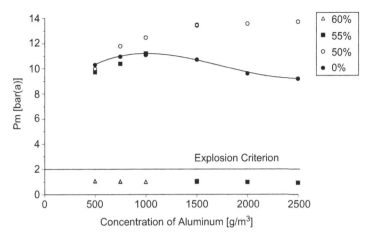

FIGURE 7.5 Explosion overpressure as a function of aluminum concentration for different amounts of admixed SBC, showing the occurrence of SEEP (Suppressant Enhanced Explosion Parameter) in a spherical 1-m³ chamber [8].

overpressure, for the case of insufficient inertant in the fuel/inertant mixture, is enhanced to levels greater than that for the pure dust. A mixture of 50% aluminum/50% SBC is seen to generate higher overpressures than the aluminum dust alone, while the addition of 5% more SBC to the mixture results in explosion inerting only at the higher aluminum dust concentrations. Once the dust mixture consists of 60% SBC, inerting is then successful at lower aluminum dust concentrations.

This behavior can be attributed to decomposition of the inertant by heat from the combustion of the fuel. For the example shown in Figure 7.5, the hot-burning aluminum powder is thought to bring about at least partial decomposition of the SBC and hence generation of carbon dioxide. Companion testing with MAP showed a similar occurrence of SEEP; this was likely caused by the generation of ammonia and hydrogen from the monoammonium phosphate [8]. An analogous situation investigated by Pegg et al. [11] is the decomposition of sodium azide (NaN_3) and subsequent generation of nitrogen leading to overpressures much greater than those possible due solely to the combustion of sodium azide. This explains the use of sodium azide as a gas generant in automobile airbags.

7.4 THERMAL INHIBITORS

Pure thermal inhibitors such as limestone and dolomite may not result in the SEEP phenomenon to an appreciable extent, but their use requires that they be added to an explosible dust in significant amounts. (Recall Figure 7.4 and the hypothetical example given in Section 4.4.) While these materials act to lower

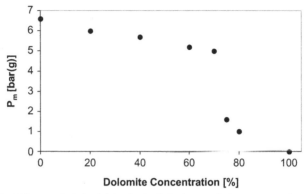

FIGURE 7.6   Influence of admixed dolomite (rock dust) on explosion pressure of a coal dust [12].

the rate of pressure rise essentially from the onset of admixture, they do not have an appreciable effect on explosion overpressure until the actual inerting level is closely approached.

This effect can be seen in Figure 7.6, which illustrates the dolomite inerting requirement of approximately 80% for a particular coal dust when tested in a spherical 26-L chamber [12]. This means that 80% of the dust mixture must be non-combustible dolomite with the remaining 20% being combustible coal dust. Expressed another way, the mass of dolomite must be four times the mass of coal dust to render the mixture inert.

7.5 REALITY

Inerting and suppression are prevention and mitigation applications, respectively, in which *a little is not good enough*. As commented by Moore [3], the consequence of insufficient deployment of a suppressant agent is a failed suppression in which the resulting overpressure can be greater than for the unsuppressed explosion (in the case of the occurrence of SEEP, or Suppressant Enhanced Explosion Parameter).

Moore [3] uses the *suppression* example of halon molecule dissociation contributing to the overall system pressure to explain the SEEP phenomenon with these materials (which are largely no longer used in new protection systems due to their ozone layer depleting feature). In addition to the aluminum/SBC and aluminum/MAP cases described in Section 7.3, other *inerting* examples in which SEEP has been observed are polyethylene/SBC and polyethylene/MAP [8], as well as coal dust/fly ash and petroleum coke dust/fly ash [13]. Even without the occurrence of SEEP, thermal inhibitors are typically required in such high percentages to be effective that their use in small amounts is of little help in protecting against overpressure.

7.6 WHAT DO *YOU* THINK?

Statements were made in Section 7.2 that limestone is (a) a thermal inhibitor and (b) less effective than sodium bicarbonate and monoammonium phosphate because it is not able to decompose in the fast-moving flame-front occurring in a dust explosion. Note that both points relate to thermal effects, with point (a) being somewhat obvious and point (b) being explained by the fact that although decomposition is a chemical reaction, it is the self-reaction of the inertant being considered, not the fuel/oxidant reaction.

On point (a), what is the physical property of limestone that makes it useful as a heat sink? (Hint: The value of this property for water is 1 cal/g·°C or 1 Btu/lb_m·°F.)

As for point (b), limestone decomposition is an endothermic reaction, meaning that heat input is required for the reaction to proceed (as opposed to an exothermic reaction such as combustion, which releases heat). So if limestone ($CaCO_3$) were to decompose, it would drain additional heat (i.e., its endothermic heat of reaction) from the reacting fuel dust and would generate a gaseous product that would also act as a thermal inhibitor. What is this gas? (Hint: See Section 2.2.)

But the decomposition of limestone does not occur at room temperature and does not occur instantaneously. What is the decomposition temperature for limestone? (An Internet search will work well here.) As for the requirement of a finite reaction time, does this help to explain the inability of limestone to decompose in the *rapidly advancing* flamefront? In other words, non-material factors such as residence time can be as important as the actual physical properties of the inertant [14].

Switching from the specific case of thermal inhibitors, we turn for a different exercise to the comparatively recent paper by Date et al. [15] These authors have moved away from a traditional deterministic approach to dust explosion protection to one that is based on assessing the residual risk of failure of explosion protection systems. This is a particularly noteworthy effort because of the recognition that even after identifying hazards, assessing risks, and selecting prevention and mitigation measures, risk still remains and this residual risk (e.g., a failed suppression) must be managed.

Figures 7.7 and 7.8 show the industrial plant used as a case study by Date et al. [15] and a directed graph representation of the plant, respectively. As an initial step, one could compare the two schematics and verify the interconnections among the plant components. Probabilities of flame propagation through these interconnections and other risk computations can be found in the full work of Date et al. [15] This reference is recommended for those readers interested in a fresh look at design decision making related to dust explosion protection.

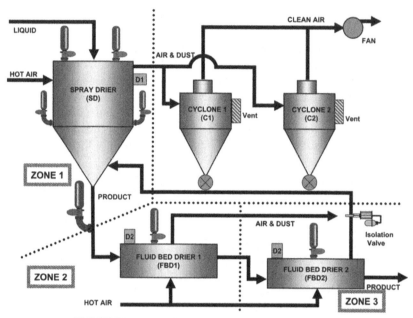

FIGURE 7.7   Spray drier plant with explosion protection [15].

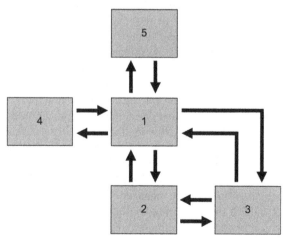

FIGURE 7.8   Directed graph representation for spray drier plant shown in Figure 7.7 [15]. Legend for components: 1—spray drier; 2—fluid bed drier #1; 3—fluid bed drier #2; 4—cyclone #1; 5—cyclone #2.

REFERENCES

[1] Klemens R. Explosions of industrial dusts—from ignition source to suppression. Krakow, Poland: Plenary Paper, Proceedings of Ninth International Symposium on Hazards, Prevention, and Mitigation of Industrial Explosions; July 22–27, 2012.

[2] Amyotte PR. Solid inertants and their use in dust explosion prevention and mitigation. Journal of Loss Prevention in the Process Industries 2006;19:161–73.

[3] Moore PE. Suppressants for the control of industrial explosions. Journal of Loss Prevention in the Process Industries 1996;9:119–23.

[4] Moore PE, Siwek R. New developments in explosion suppression. VDI Berichte 1992;975:481–505.

[5] Dastidar A, Amyotte P. Determination of minimum inerting concentrations for combustible dusts in a laboratory-scale chamber. Process Safety and Environmental Protection 2002;80:287–97.

[6] Eckhoff RK. Dust explosions in the process industries, 3rd ed. Boston, MA: Gulf Professional Publishing/Elsevier; 2003.

[7] Chatrathi K, Going J. Dust deflagration extinction. Process Safety Progress 2000;19:146–53.

[8] Dastidar AG, Amyotte PR, Going J, Chatrathi K. Flammability limits of dusts—minimum inerting concentrations. Process Safety Progress 1999;18:56–63.

[9] Dastidar AG, Amyotte PR, Going J, Chatrathi K. Scaling of dust explosion inerting. Archivum Combustionis 1998;18:21–45.

[10] Amrogowicz J, Kordylewski W. Effectiveness of dust explosion suppression by carbonates and phosphates. Combustion and Flame 1991;85:520–2.

[11] Pegg MJ, Amyotte PR, Lightfoot PD, Lee MC. Dust explosibility characteristics of azide-based gas generants. Journal of Loss Prevention in the Process Industries 1997;10:101–11.

[12] Amyotte PR, Mintz KJ, Pegg MJ, Sun YH, Wilkie KI. Effects of methane admixture, particle size and volatile content on the dolomite inerting requirements of coal dust. Journal of Hazardous Materials 1991;27:187–203.

[13] Amyotte PR, Khan FI, Basu A, Dastidar AG, Dumeah RK, Erving WL. Explosibility parameters for mixtures of pulverized fuel and ash. Krakow, Poland: Proceedings of Fifth International Symposium on Hazards, Prevention, and Mitigation of Industrial Explosions; October 10–14, 2004.

[14] Amyotte PR, Mintz KJ, Pegg MJ. Effectiveness of various rock dusts as agents of coal dust inerting. Journal of Loss Prevention in the Process Industries 1992;5:196–9.

[15] Date P, Lade RJ, Mitra G, Moore PE. Modelling the risk of failure in explosion protection systems. Journal of Loss Prevention in the Process Industries 2009;22:492–8.

Myth No. 7 (Ignition Source): Dusts Ignite Only with a High-Energy Ignition Source

In the introductory paragraphs of Chapter 5, possible differences in ignition requirements between flammable gases and combustible dusts were removed from the discussion by use of the example of an open flame. Such an ignition source would intuitively be sufficient to initiate combustion of either gaseous or particulate fuels. Here, in Chapter 8, we consider how the typically low ignition energy requirements of gases may lead to the misconception that all dusts require correspondingly higher energies for ignition.

Thus arises the myth that a dust explosion can be initiated only with the aid of a high-energy ignition source. Again we see the need to distinguish between hazard and risk when considering the relationship of dust explosibility parameters to actual processing conditions. In a sense, this is a subset of the Chapter 5 myth that gas explosions are much worse than dust explosions. (See especially Section 5.2.)

8.1 INDUSTRIAL IGNITION SOURCES

In their comprehensive review of the causes, consequences, and control of dust explosions, Abbasi and Abbasi [1] list the following *dust explosion triggers*:

- Flames and direct heat
- Hot work (e.g., welding and cutting)
- Incandescent material (e.g., smoldering particles)
- Hot surfaces, (e.g., overheated bearings)
- Electrostatic sparks (caused by electrostatic discharge from electrical equipment)
- Electrical sparks (such as may be caused by switching operations)
- Friction sparks and hot spots (caused by rubbing between solids and friction-induced heating, respectively)
- Impact sparks (ignition by surface heating resulting from metal-on-metal impact)
- Self-heating (spontaneous combustion)
- Static electricity (electrostatic sparks caused by process operations such as pouring and size reduction)
- Lightning
- Shock waves.

Abbasi and Abbasi [1] comment that with the preceding triggers differing in origin and nature (temperature, energy, etc.), it is clear that dusts can be ignited by low-energy as well as high-energy sources. This is consistent with the theme of the current chapter, as is the categorization given by Taveau [2] of ignition sources for dust explosions occurring in Germany during the period 1965–1985:

- Mechanical sparks
- Smoldering nests
- Mechanical heating and friction
- Electrostatic discharges
- Fire

- Spontaneous ignition (self-ignition)
- Hot surfaces
- Welding and cutting
- Electrical machinery
- Unknown or not reported
- Others.

Although Taveau [2] uses slightly different terminology, his preceding listing is similar to that given by Abbasi and Abbasi [1]. (For example, ignition by mechanical heating [2] is equivalent to hot-surface ignition [1].) Taveau [2] further introduces the notion that it is not unusual for an incident investigation to conclude that the precise ignition source could not be determined. The reason could be due to the complexity of the scenario in terms of level of destruction, or perhaps because of the abundance of potential ignition sources that could have been involved.

8.2 STANDARDIZED DUST EXPLOSIBILITY TESTING

Many of the ignition sources listed in the preceding section involve an elevated-temperature surface or an energetic spark. To relate these process features to a specific material hazard, testing using standardized equipment and procedures is conducted (as previously mentioned in Section 2.2). Examples include the MIKE 3 apparatus (see Figure 8.1) and ASTM E2019 [3] for determination of dust cloud minimum ignition energy (MIE), and the BAM oven (see Figure 8.2) and ASTM E1491 [4] for determination of dust cloud minimum ignition temperature (MIT).

Note: As indicated in the captions for Figures 8.1 and 8.2, the testing equipment shown was manufactured by Kuhner AG in Basel, Switzerland. Other manufacturers of similar apparatus do exist, and some testing/research organizations have sufficient in-house capabilities to produce their own devices. Further, in addition to the referenced ASTM standards widely used in North America, testing standards for dust explosibility parameters do exist in other parts of the world (European, Asian, and other international standards). Reference to Kuhner apparatus and ASTM standards in this book is driven by purely pragmatic reasons; these have been employed in our laboratory at Dalhousie University to good effect over the past two decades and are therefore well known to me.

The MIKE 3 apparatus (Figure 8.1) consists of a 1.2-L cylindrical glass chamber into which dust is dispersed and then ignited by an electrical spark of known energy (maximum value of 1000 mJ, or 1 J). Dust dispersal occurs by means of an air blast raising the sample previously placed at the base of the glass tube into suspension. The ignition criterion is flame propagation at least 6 cm away from the spark electrodes (shown in Figure 8.1 about one-third of the way up the tube from its base). This usually results in flame exiting the top of the tube, which is covered by a hinged metal plate.

FIGURE 8.1   MIKE 3 apparatus for determination of minimum ignition energy, or MIE (manufactured by Kuhner AG, Basel, Switzerland).

The BAM oven (Figure 8.2) is a cylindrical furnace named for the initials of the federal research institute in Germany where it was first developed. A dust cloud is generated by squeezing a rubber bulb previously charged with sample (shown on the right in Figure 8.2), which directs the dust against a circular, concave metal plate of about 20 cm^2 area and known temperature (maximum value of 600 °C). Dust ignition is assessed by visual observation of flame exiting the back end of the oven, which is covered by a hinged metal flap (similar to the MIKE 3 apparatus).

Chapter 19 discusses at length the matter of standardized testing for the parameters P_{max}, K_{St}, and MEC, which are determined in closed vessels so that explosion pressure is not relieved during the test sequence as with the MIKE 3 apparatus and

FIGURE 8.2 BAM oven for determination of minimum ignition temperature, or MIT (manufactured by Kuhner AG, Basel, Switzerland).

BAM oven. A high dispersing air pressure is used in these closed-chamber tests, and this means that a correspondingly high turbulence intensity exists in the test vessel prior to dust ignition. As a consequence, highly energetic chemical ignitors having a stored energy of 10 kJ are used to determine parameters such as P_{max} and K_{St}.

Again, these considerations are covered further in Chapter 19; Chapter 14 also deals with the influence of turbulence on dust explosibility. Mention is made of these points here because it is likely that these applied energies in the kilojoule range help give the impression of dusts being ignitable only with high-energy triggers. It must be remembered, however, that when energetic sources are used in closed-vessel dust explosibility testing, the reason is to *ensure* ignitability given the highly turbulent pre-ignition dust cloud conditions. When the intent is to *measure* ignitability, lower dispersing air pressures are used (as in the MIKE 3 apparatus and BAM oven), and ignition becomes possible with lower energies depending on the nature of the combustible material.

The next section discusses just how low these energies—particularly spark energies—can be for dusts. First, though, mention is made of a unique aspect of spark-generating circuit design that can affect the values measured for minimum ignition energy—i.e., whether or not provision for inductance has been included in the discharge circuit.

Quoting from the MIKE 3 apparatus manual [5] (p. 1) provided by the manufacturer:

By definition, minimum ignition energy data refer to protracted capacitor discharges. These are generally clearly more incendive than purely capacitive discharges. The results obtained under such conditions can be applied to operational conditions only if the capacitors occurring in plant installations are also discharged via an inductance. Hence, if the incendivity of electrical discharges—especially of electrostatic discharges—with regard to dust/air mixtures is to be assessed, the minimum ignition energy must also be determined without an inductance in the discharge circuit.

Thus, the use of inductance—1 mH (millihenry) in the case of the MIKE 3 apparatus—results in a longer duration (i.e., protracted) spark with the potential for a lower MIE than would be measured using no inductance in the spark circuitry. As noted in ASTM E2019 [3], almost all electrostatic discharges in plant installations are capacitive with negligible inductance. Practical evidence indicates that most electrostatic ignition sources have energies less than 1 J and that common electrostatic ignition energies are typically less than 30 mJ (Dastidar, A.G., 2012. Personal communication, with permission).

Table 8.1 illustrates the influence of inductance on reported MIE values for selected materials; in the majority of cases, the use of inductance leads to a lower (sometimes significantly) MIE. The effect is not entirely straightforward, but varies and depends on the dust type [5]. For those dusts that experience a lowering of the MIE with added inductance, von Pidoll [6] offers a plausible explanation based on the underlying physics. He reasons that because purely capacitive sparks generate stronger shock waves than time-delayed inductive sparks, separation of the dust particles from the ignition source may be more likely with the former spark type [6]. This would enhance the ignition probability of inductive sparks and possibly lower the MIE.

TABLE 8.1 Comparison of Minimum Ignition Energy Values With and Without Inductance

Material	MIE with inductance [mJ]	MIE without inductance [mJ]
Organic stabilizer	0.4	0.4
Benzanthron	0.9	1.0
Epoxy coating powder	1.7	2.5
Polyurethane coating powder	2	8
Epoxy-polyester coating powder	2.3	9
Polyester coating powder	2.9	15
Polyamide coating powder	4	19
Lycopodium	5	50
Magnesium granulate	25	200
Aluminum granulate	50	500
Flock	69–98	1300–1600

(adapted from von Pidoll [6])

Given the previous discussion in Chapter 6 on the desirability of reporting particle size distributions, you may be looking for these data with respect to the MIE values in Table 8.1. That is an excellent observation and a reasonable expectation. As with the original reference [6] for Table 8.1, the intent here is to focus on the effect of inductance, so a full reporting of material properties is not provided. By necessity, this will occur at various times throughout the current book.

8.3 DUST CLOUD IGNITION BY LOW-ENERGY SOURCES

We begin our look at dust cloud ignition by low-energy sources with the following excerpt from the 2001 paper by von Pidoll [6] (p. 107):

In comparison to gases and vapors which usually have MIEs between 0.02 mJ and 0.5 mJ, solid particles have a higher MIE in the range from 1 mJ to 1000 mJ. Due to the high ignition energy needed, the destruction capability of a dust cloud explosion is often underestimated.

This passage speaks of the origin of the notion that dusts require high-energy ignition sources. MIEs for dusts may in general be somewhat higher than for gases, but should a dust MIE of 1 mJ (or as we shall soon see, < 1 mJ) be considered a *high* value of this parameter? And even if one were to say that a given dust MIE is high in comparison to a reference gas MIE, what if process conditions are such that there exist actual ignition energies orders of magnitude greater than the dust MIE? What is the comparison that should be made in this case? (Recall the discussion in Chapter 5.)

Minimum ignition energy determination for dusts is an active area of research in the scientific and engineering literature, as evidenced by a listing of some of the papers published over the last decade or so [6–16]. Professor Rolf Eckhoff of the University of Bergen, Norway, has presented and published extensively on the topic of spark ignition of dusts. In a 2007 paper [9], he and his colleague remarked that until about 1975, common wisdom held that all dusts had MIEs greater than 10 mJ. As progress in the development of spark generation apparatus made lower-energy sparks attainable for dust studies, MIEs of 1 mJ and < 1 mJ began to be reported with increasing regularity [9,10].

Why then is it still difficult in some quarters to believe that such low-energy sources can lead to full-scale dust explosions—notwithstanding the previously mentioned factors of high-energy ignitors being used in closed-vessel explosion testing and dust MIEs generally being greater than gas MIEs?

I personally believe an additional reason is that the field of electrostatic charging and generation of electrical sparks is foreign to many people working in the process industries. I know this is indeed the case for me as a chemical engineer by education, training, and practice. So while I am comfortable with concepts such as solvent vapor pressure and how this property determines vapor concentration at different temperatures, I am less familiar and comfortable with the physics of electrical charges and discharges. Speaking with colleagues of similar background in industry, I sense that I am not alone. What works well for me in these situations is to first read and learn from a well-written, fundamentals-based paper prepared for the generalist technical audience before delving into the literature written more for specialists. I also look for papers of this nature that have been published by authoritative, respected organizations.

The 1994 paper "Tame Static Electricity" by Louvar et al. [17], published in *Chemical Engineering Progress* by the American Institute of Chemical Engineers, has been quite helpful to me in attempting to understand the basics of the subject matter. In particular, the authors show the general relationship between energies from electrostatic discharges of various types and the energies required to ignite flammable gases and combustible dusts. The nature of the different discharge mechanisms is also well explained by Louvar et al. [17]. The paper by Glor [18], published in 2010 in *Pharmaceutical Engineering* by the International Society for Pharmaceutical Engineering, also gives a good description of typical discharges that could occur during bulk powder transfers.

A recent (2011) article by Pavey [19], published in *The Chemical Engineer* by the Institution of Chemical Engineers, provides a good explanation of the primary static charge generation mechanisms of tribo-charging and induction charging. Pavey [19] also discusses static discharges and therefore gives an update on some of the features presented by Louvar et al. [17]. For example, Louvar et al. [17] identify dusts as not being ignited by corona-type discharges and unlikely to be ignited by brush-type discharges. The following excerpts from Pavey [19] (p. 29) show that current evidence supports these earlier findings:

Brush discharges occur from insulating surfaces. The energy in a brush discharge is relatively low, often below the limit of human perception. Nevertheless, brush discharges are quite capable of igniting most common flammable vapours and gases. Ignition of a dust cloud with a brush discharge has never been positively confirmed even under laboratory conditions, but it is still often considered prudent to assume it might be possible for the most sensitive of powders.

and

Corona discharges are low-energy discharges from sharp points, and are unlikely to ignite any but the most sensitive of gases, such as hydrogen.

To conclude this section on low-energy ignition of dust clouds, we briefly examine two industrial case studies and some recent research results for nano-powders. The first case history is that presented by Kao and Duh [20] involving a series of dust explosions in the silo area of an ABS (acrylonitrile-butadiene-styrene) plant; the participating materials were ABS (which is a rubber-containing plastic) and SAN (polystyrene-acrylonitrile). Several silos were affected, with the top-plate and bag-filter (dust collector) for each being destroyed. An indication of the plant damage is given by Figure 8.3.

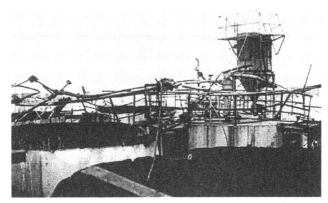

FIGURE 8.3 Top view of ABS (acrylonitrile-butadiene-styrene) silo after explosion [20].

The incident investigation team eventually concluded that the explosion was initiated in one of the silos undergoing gravitational filling of ABS powder, with flame propagation to the other involved silos occurring through interconnecting pipes. The most likely ignition scenario in the first silo was thought to be a bulked brush (conical pile or cone) discharge between the compacted powder and the grounded silo wall [20]. Such discharges are described by Louvar et al. [17] as being "relatively intense" and having energies up to several hundred millijoules. The MIE of the ABS powder was measured as 10 mJ [20].

A second example presented by Matsuda and Yamaguma [21] describes a tantalum (metal) powder explosion in a bag-filter dust collector. There are at least two significant lessons afforded by this unfortunate incident, in which casualties occurred. First, the presence of a thin, non-conductive, and high-resistivity oxide layer on the powder surface was determined to have contributed to static charge generation when the powder was flowing through the dust collector (thus highlighting the importance of changes in material characteristics). The MIE of the tantalum was determined in laboratory testing to be 14 mJ; this energy was apparently exceeded by the electrostatic discharge postulated to have occurred between the charged tantalum and the polyethylene bag lining the dust collector [21].

A second important feature of this case is the process of elimination used by the authors to determine a plausible ignition scenario. Matsuda and Yamaguma [21] describe their efforts to investigate the possibility of ignition by mechanical sparks, external sources such as flames, and self-heating. This process eventually led to the conclusion of electrostatic spark ignition, thus reinforcing the need for a comprehensive prevention program aimed at all potential ignition sources. Although not necessarily the case here, one is also reminded of the fact that the assessment of process risk from unidentified hazards is impossible.

Finally, a new area of dust explosion concern has emerged in recent years on a very small scale—at least with respect to particle size. Nano-size powders are becoming increasingly important in industry, and the question of particle size

TABLE 8.2 Ignitability Parameters for Micron- and Nano-Size Titanium

Size	MIE [mJ] With inductance (E_s)	MIE [mJ] Without inductance (E_s)	MIT [°C]
–100 mesh (< 150 μm)	10–30 (23)	1–3 (1.7)	> 590
–325 mesh (< 45 μm)	1–3 (1.7)	1–3 (2.3)	460
≤ 20 μm	< 1	< 1	460
150 nm	Not Determined	< 1	250
60–80 nm	Not Determined	< 1	240
40–60 nm	Not Determined	< 1	250

(adapted from Amyotte et al. [25])

influence on explosibility parameters for these materials is being aggressively pursued in several laboratories throughout the world. Eckhoff [22] gives an excellent and timely summary of this work, and provides solid fundamental reasoning for the results obtained to date. Although explosion severity (as measured by P_{max} and K_{St}) for nano-dusts does not appear to be significantly greater than for micron-size dusts, the same cannot be said for ignitability parameters. This is especially the case for nano-size metals such as iron [23] and titanium [23,24], which have been shown experimentally to exhibit very low spark ignition energies (< 1 mJ).

Ignitability data for micron- and nano-size titanium samples are shown in Table 8.2 as determined using a MIKE 3 apparatus for MIE and a BAM oven for MIT [25]. MIE results are reported as a range of values both with and without an inductance of 1 mH; this is an artifact of the MIKE 3 apparatus, which uses set ignition energies of 1, 3, 10, 30, 100, 300, and 1000 mJ. Where applicable, E_s (the "statistic energy" determined by manufacturer [Kuhner]-supplied software based on probabilistic considerations) is also given in parentheses after the range.

The MIT data in Table 8.2 illustrate that nano-size titanium is ignitable by relatively low hot-surface temperatures (much lower than for the micron-size samples). Both micron- and nano-size titanium are ignitable by low spark energies, as shown by the MIE data in Table 8.2.

The enhanced spark ignitability of the nano-titanium is illustrated by Figure 8.4, which compares MIKE 3 ignition graphs for the –325 mesh (< 45 μm) and 150 nm samples, respectively in (A) and (B). (In Figure 8.4, as with Figure 5.2, the dust amount appears on the x-axis and spark energy on the y-axis; the open boxes indicate no ignition at the particular delay time, and the solid boxes indicate ignition.) The nano-size sample is clearly ignitable at lower spark energies and significantly lower dust amounts (concentrations) than is the micron-size sample.

The ignitability enhancement with a decrease in particle size into the nanometer range is even more striking for iron dust, as demonstrated by Wu et al. [23].

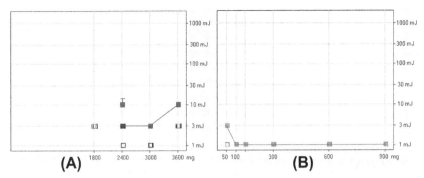

FIGURE 8.4   (A) Minimum ignition energy data (without inductance) for –325 mesh (< 45 μm) titanium. (B) Minimum ignition energy data (without inductance) for 150 nm titanium [25].

These authors determined that a 150-μm average diameter iron sample (range of 75–225 μm) would not ignite in a MIKE 3 apparatus, whereas all three nanometer-size iron samples tested had MIE values < 1 mJ. Findings such as these have led Eckhoff [22] (p. 458) to conclude that consideration must be given to (with underlining added for emphasis):

the possibility of ignition by <u>subtle</u> electrostatic spark discharges of <u>very low</u> energies

and development of

a new MIE test apparatus capable of producing synchronized electrical spark energies <u><< 1 mJ</u>.

8.4   REALITY

Compelling evidence exists that under the intentionally low turbulence levels present in standardized MIE test apparatus, some dusts will ignite at spark energies as low as 1 mJ. Ignition energies in the kilojoule range, although routinely used in laboratory closed-vessel testing conducted at high turbulence intensities, are not usually representative of the actual minimum energy required for ignition of dusts.

As seen in previous chapters, the fundamental nature of a particulate fuel means that it may well be harder to ignite than a gaseous counterpart and may react more slowly once ignited. The only comparison of real value, however, is how the required ignition energy and ignition temperature of a given dust relate to the available energies and temperatures it will experience under normal and upset process conditions. Basic treatments of ignition parameters written for non-specialists can be helpful in making such a comparison.

Ignition sources abound in industry. As more than one industrial colleague has said to me on more than one occasion—if there is one thing that is free in industry, it is an ignition source. And it seems that the list of available ignition sources continues to grow.

Recent work by Simon et al. [26] has shown that under certain conditions it is possible to ignite dust/air mixtures with ultrasonic waves. They conducted tests with sulfur, calcium stearate, and maize starch using an apparatus similar to the MIKE 3 shown in Figure 8.1, but with the electrodes replaced by a high-absorption coefficient target material that was heated by acoustic waves from an ultrasound-generating unit. The overall conclusion was that it would be unlikely for all the necessary conditions for ultrasound ignition to be met at once accidentally [26]. This may indeed be the case, but a cautionary note should still be sounded.

8.5 WHAT DO *YOU* THINK?

The key points of discussion in this chapter have been (i) the wide variety of ignition sources typically found in an industrial setting (or which have the potential to be introduced through human action or inaction) and (ii) the fact that many of these are in the low-millijoule energy range. How then, in your facility, do you account for these points?

Clearly, ignition sources need to be avoided to the greatest extent possible to facilitate effective explosion prevention; yet potential ignition sources seem to be an inevitable occurrence in industrial processing of powders. So perhaps a more pertinent question is—how much do you rely on ignition source removal as your primary line of defense against dust explosions? In a similar vein, what emphasis do you place on the avoidance of combustible dust cloud formation? Or on the removal of the oxygen required for combustion? What are the implications of your answers to these questions with respect to explosion mitigation measures to protect plant personnel and equipment?

When you are considering the preceding questions, the excerpted comments from Glor [18] given in Section 5.4 can provide guidance in operations where electrostatic issues are prevalent, as in the case of combustible powder transfer to a process vessel (possibly containing a flammable solvent). Expressed differently, again in the words of Glor [18] (p. 1):

...this type of operation is clearly one of the most hazardous within the process industry if exclusion of effective ignition sources is the only basis of safety. If all those effective ignition sources generally considered common..., including those ignition sources related to electrical equipment, mechanical load, open flames, cutting, welding and smoking, etc., have been excluded through the introduction of precautionary measures, the hazard of electrostatic ignition inherent in the powder transfer remains a viable possibility for causing an explosion.

As detailed in Section 8.2, one of the requirements in assessing explosion likelihood is to relate MIE results from standardized tests to the actual industrial setting. An important consideration in this regard is whether or not the MIE was determined with inductance in the spark generation circuitry. To recap, inductance produces sparks that are more incendive (i.e., able to cause ignition) than those resulting from circuits without inductance, thus generally resulting in a

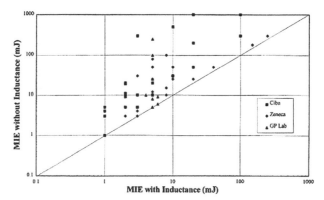

FIGURE 8.5 Influence of inductance on minimum ignition energy [27]. The names given in the data legend are those of the organizations that conducted the testing.

more conservative measurement of MIE (Dastidar, A.G., 2012. Personal communication, with permission). MIE tests without inductance typically better simulate sparks from pure/ideal electrostatic sources, whereas tests with inductance better simulate sparks from electrical/electronic sources (Dastidar, A.G., 2012. Personal communication, with permission).

Figure 8.5 shows the influence of inductance on selected MIE values. Is this effect consistent with the discussion in Section 8.2? Were all the tested dusts affected in the same manner by the use of inductance in the spark circuit? (Note that the diagonal in Figure 8.5 represents the case of identical MIE values with and without inductance.)

REFERENCES

[1] Abbasi T, Abbasi SA. Dust explosions—cases, causes, consequences, and control. Journal of Hazardous Materials 2007;140:7–44.

[2] Taveau J. Secondary dust explosions: how to prevent them or mitigate their effects? Process Safety Progress 2012;31:36–50.

[3] ASTM. ASTM E2019–03, Standard test method for minimum ignition energy of a dust cloud in air. West Conshohocken, PA: American Society for Testing and Materials; 2003.

[4] ASTM. ASTM E1491–06, Standard test method for minimum autoignition temperature of dust clouds. West Conshohocken, PA: American Society for Testing and Materials; 2006.

[5] Kuhner. MIKE 3 Manual. Kuhner AG. Basel: Switzerland; 2012. Available at http://safety.kuhner.com/tl_files/kuhner/product/safety/PDF/B021_071.pdf (last accessed October 7, 2012).

[6] von Pidoll U. The ignition of clouds of sprays, powders and fibers by flames and electric sparks. Journal of Loss Prevention in the Process Industries 2001;14:103–9.

[7] Eckhoff RK. Minimum ignition energy (MIE)—a basic ignition sensitivity parameter in design of intrinsically safe electrical apparatus for explosive dust clouds. Journal of Loss Prevention in the Process Industries 2002;15:305–10.

[8] Randeberg E, Olsen W, Eckhoff RK. A new method for generation of synchronised capacitive sparks of low energy. Journal of Electrostatics 2006;64:263–72.

[9] Randeberg E, Eckhoff RK. Measurement of minimum ignition energies of dust clouds in the < 1 mJ region. Journal of Hazardous Materials 2007;140:237–44.

[10] Eckhoff RK, Randeberg E. Electrostatic spark ignition of sensitive dust clouds of MIE < 1 mJ. Journal of Loss Prevention in the Process Industries 2007;20:396–401.

[11] Janes A, Chaineaux J, Carson D, Le Lore PA. MIKE 3 versus Hartmann apparatus: comparison of measured minimum ignition energy (MIE). Journal of Hazardous Materials 2008;152:32–9.

[12] Marmo L, Cavallero D. Minimum ignition energy of nylon fibres. Journal of Loss Prevention in the Process Industries 2008;21:512–17.

[13] Lu K-T, Chen T-C, Hu K-H. Investigation of the decomposition reaction and dust explosion characteristics of crystalline benzoyl peroxides. Journal of Hazardous Materials 2009;161:246–56.

[14] Lu K-T, Chu Y-C, Chen T-C, Hu K-H. Investigation of the decomposition reaction and dust explosion characteristics of crystalline dicumyl peroxides. Process Safety and Environmental Protection 2010;88:356–65.

[15] Choi K, Sakurau N, Yanagida K, Itoh H. Ignitability of aluminous coating powders due to electrostatic spark. Journal of Loss Prevention in the Process Industries 2010;23:183–5.

[16] Bernard S, Gillard P, Foucher F, Mounaim-Rousselle C. MIE and flame velocity of partially oxidised aluminum dust. Journal of Loss Prevention in the Process Industries 2012;25:460–6.

[17] Louvar JF, Maurer B, Boicourt GW. Tame static electricity. Chemical Engineering Progress 1994;90:75–81.

[18] Glor M. A synopsis of explosion hazards during the transfer of powders into flammable solvents and explosion preventative measures. Pharmaceutical Engineering 2010;30:1–8.

[19] Pavey I. No charge please. The Chemical Engineer 2011;838:28–30.

[20] Kao C-S, Duh Y-S. Accident investigation of an ABS plant. Journal of Loss Prevention in the Process Industries 2002;15:223–32.

[21] Matsuda T, Yamaguma M. Tantalum dust deflagration in a bag filter dust-collecting device. Journal of Hazardous Materials 2000;A77:33–42.

[22] Eckhoff RK. Does the dust explosion risk increase when moving from μm-particle powders to powders of nm-particles? Journal of Loss Prevention in the Process Industries 2012;25:448–59.

[23] Wu H-C, Chang R-C, Hsiao H-C. Research of minimum ignition energy for nano titanium powder and nano iron powder. Journal of Loss Prevention in the Process Industries 2009;22:21–4.

[24] Boilard SP, Amyotte PR, Khan FI, Dastidar AG, Eckhoff RK. Explosibility of micron- and nano-size titanium powders. Krakow, Poland: Proceedings of Ninth International Symposium on Hazards, Prevention, and Mitigation of Industrial Explosions; July 22–27, 2012.

[25] Amyotte P, Khan F, Boilard S, Iarossi I, Cloney C, Dastidar A, Eckhoff R, Marmo L, Ripley R. Explosibility of non-traditional dusts: experimental and modeling challenges. Southport, UK: Hazards XXIII, IChemE Symposium Series No. 158; Nov. 13–15, 2012. pp. 83–90.

[26] Simon LH, Wilkens V, Fedtke T, Beyer M. Ignition of dust-air atmospheres by ultrasonic waves. Krakow, Poland: Proceedings of Ninth International Symposium on Hazards, Prevention, and Mitigation of Industrial Explosions; July 22–27, 2012.

[27] Bailey M, Hooker P, Caine P, Gibson N. Incendivity of electrostatic discharges in dust clouds: the minimum ignition energy problem. Journal of Loss Prevention in the Process Industries 2001;14:99–101.

Chapter 9

Myth No. 8 (Ignition Source): Only Dust Clouds—Not Dust Layers—Will Ignite

Layered dust does not explode. This fact has been emphasized time and again in previous chapters when discussing, for example, the explosion pentagon or the relatively thin layers required to form an explosible dust cloud once dispersed. Further, this is a book on dust explosions, not dust fires. So why include a chapter on dust layer ignition?

The reason is quite simple. If we completely ignore dust layer ignition and fires when addressing dust explosions, then do we not contribute to the myth that only dust clouds (and therefore not dust layers) are ignitable?

It does seem to me that the worlds of dust explosions and dust fires do not often mix in the technical literature. Certainly, my own expertise lies almost exclusively in the field of dust explosions, with perhaps just one exception [1]. Nevertheless, I consider it important to examine, even briefly, the topic of ignition of dust layers. For layered dust does not always become suspended in air and explode—sometimes it catches fire.

In their extensive treatments of dust explosions, both Barton [2] and Eckhoff [3] have taken a similar approach to that adopted here by including some discussion on the ignition and flammability of non-airborne dust. See, for example, pages 16–17 and 261–263 in Barton [2], and pages 501–507 and 564–565 in Eckhoff [3].

9.1 DUST LAYER IGNITION

The words *self, auto,* and *spontaneous* are often used in front of the words *heating, ignition,* and *combustion* to form various combinations (e.g., *self-ignition* and *spontaneous combustion*). One also sees reference in the literature to dust *layers, piles,* and *deposits.* Sometimes these terms appear to be used interchangeably—but is there a difference in meaning?

My general sense is that piles and deposits are similar and are deeper (or thicker) than layers. And self- or spontaneous heating, ignition, or combustion occurs without an external ignition source such as a hot surface (e.g., an overheated bearing). At least that is the usage I generally attempt to follow, and is an approach that is consistent with the paper titles in the reference list for this chapter. It is also consistent with the terminology used in, again, the authoritative texts by Barton [2] (p. 261):

Wherever dusts are handled, layers or deeper deposits may form. Fires may start in a layer or deposit, most commonly by hot surfaces or spontaneous heating.

and Eckhoff [3] (p. 501) in a section heading:

Ignition of dust deposits/layers by self-heating or hot surfaces.

Figure 9.1 illustrates the process of self-heating in a coal pile with eventual self- (or spontaneous) ignition and combustion. The stages involved in this process are well represented by the requirements for (i) a degree of porosity sufficient for air (oxidant) to enter the pile and initiate self-heating via chemical reaction

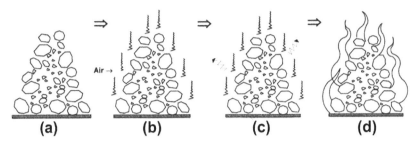

(a)   (b)   (c)   (d)

FIGURE 9.1   Evolution of a coal pile fire: (a) coal pile having various particle sizes; (b) air ingress and initial self-heating; (c) heat losses by conduction, convection, and radiation at rates not exceeding the rate of heat generation; (d) self-ignition at some critical temperature [4].

with the available fuel; (ii) the ensuing generation of heat within the pile at a rate faster than the loss of heat by the mechanisms of conduction within the pile, convection from the pile surface to the atmosphere, and radiation from the pile surface; and (iii) the attainment of a critical temperature at which thermal runaway and ignition occur [4,5].

The process of heating and oxidation of particulate matter can, of course, be aided by an external heat source. Eckhoff [3] gives two possible industrial scenarios for a hot object being in close proximity to a combustible dust: (i) a piece of metal (e.g., a nut or bolt) present in a bulk powder stream and which has been heated by repeated contact with the process equipment boundaries, and (ii) an overheated surface (e.g., a bearing or motor) covered with a layer of dust.

The latter case presented here gives rise to the flammability parameter known as the layer ignition temperature, or LIT. (See Table 5.1.) This parameter is the subject of ongoing research from a number of perspectives including experimentation [6–12], numerical modeling [13], the co-presence of inert additives with a combustible dust [1,14,15], and the co-presence of combustible additives with a combustible dust [16].

Barton [2] cautions against viewing the LIT as a "universal value" for assessing the hazards of dust accumulations. He remarks that both the geometry (e.g., thickness) and state (e.g., forced convection currents) of a given accumulation can affect the measured layer ignition temperature. Layer geometry effects are demonstrated in Figures 9.2 and 9.3, which illustrate the influence of layer thickness and diameter, respectively, on the LIT of various dust samples. Results such as these provide empirical evidence for the ignition temperature of a thin dust layer likely being higher than that of a bulk deposit of the same material [2].

9.2 DUST LAYER FIRES

How common are dust layer fires initiated by self-heating, hot surfaces, or one of the other applicable ignition sources listed in Section 8.1? This is a difficult

An Introduction to Dust Explosions

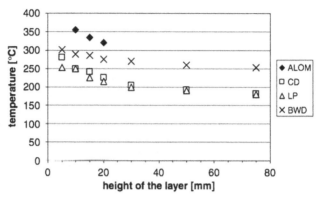

FIGURE 9.2 Effect of dust layer thickness on layer ignition temperature (LIT) [8]. Materials tested were aluminum oxide (ALOM), coal dust (CD), lycopodium (LP), and beechwood dust (BWD).

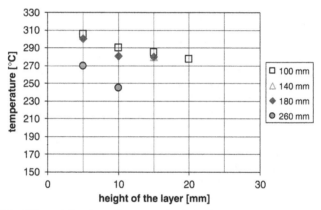

FIGURE 9.3 Effect of dust layer diameter (100, 140, 180, and 260 mm) on layer ignition temperature (LIT) [8]. Hot surface diameter was 263 mm.

question to answer given that reports of combustible dust incidents usually do not distinguish between those involving fire only, explosion only, or both (and in which sequence). Further, database information is generally not available for combustible dust near-misses and incidents involving no significant personnel harm or property damage [17]. One estimate based on research of media accounts places the majority of combustible dust incidents in the fire category, usually with no fatalities or serious injuries and only minor damage to physical assets [18].

As with any hazardous occurrence that results in little or no loss to people and property, there may be a tendency to essentially dismiss the predictive nature of dust layer fires that are detected and extinguished at an early stage. This is especially worrisome if such fires are a recurring feature of operations in a given facility. Astad [18] calls this *normalization of deviation*, whereby repeated dust fires are not viewed as possible precursors to more devastating events, but rather become accepted as a normal part of routine operation.

The act of accepting as normal—and then ignoring—such negative events has also been termed *normalizing the evidence* [19], *normalization of risk* [20], and *normalization of deviance* [21]. This practice is a key component of a culture of risk denial [19] and is therefore counter to the establishment of a well-functioning safety culture. (See Chapter 21 for further discussion of safety culture in relation to dust explosion incidents.)

Hopkins [19] uses the example of the 1996 Gretley coal mine incident in New South Wales, Australia, to illustrate how evidence normalization can have disastrous results. Repeated reports of water leaking out of the Gretley mine face were dismissed on the basis that this occurrence was expected given the generally wet state of the mine. Four miners later drowned when mining operations led to a breakthrough into an old work area that had been previously flooded. The reports of mine-face seepage had been normalized and had therefore lost their significance as warning signs.

In a similar vein, a dust layer fire should be interpreted as evidence that something is not right; there is nothing *normal* about an unintentional dust fire. At a minimum, the dust being handled has been shown to be combustible and housekeeping efforts have been shown to be inadequate on the particular occasion. If dust layer fires are a regular occurrence, then something more systemic is not right. There are likely deeper deficiencies in the safety management system, the safety culture, or both. This is a theme we will see again in Chapter 21 in our examination of the myth: *It won't happen to me*. I had originally intended the *it* in this myth to be a dust explosion; there is no reason *it* cannot also be a dust layer fire.

9.3 REALITY

We have seen previously in this book that there is an ever-present threat of layered dust being swept into the air to form a dust cloud that subsequently ignites. This is the genesis of a dust explosion. In the current chapter, we have seen that such layers (or deeper piles) can themselves ignite due to self-heating or an external hot surface either conveyed into the dust deposit or on which the deposit rests. This is the genesis of a dust fire.

I commented in Section 1.3 that the writing of this book has been motivated by a desire to aid in the protection of people, business assets, operational production, and the natural environment. This is an expression of the modern integrated approach to process safety in which all forms of loss are addressed—people, property, process, and environment. Although not the primary focus of the chapters presented here, it should be apparent that loss in all four of these areas can result from dust fires. The visual presentation in Figure 9.4 of a magnesium dust layer fire helps to confirm the validity of this statement. Clearly, such a dust fire has the capacity to cause damage in

(a)

(b)

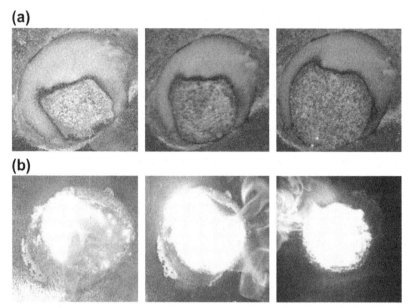

FIGURE 9.4   Burning of a magnesium dust layer: (a) first stage involving slow, lower temperature combustion; (b) second stage involving rapid, higher temperature combustion of magnesium gas [10].

and of itself, or as the precursor to a dust explosion in terms of acting as an ignition source.

9.4 WHAT DO *YOU* THINK?

A recent newspaper article [22] carried the headline "Fire Caused by Spontaneous Combustion," and contained the following passage:

Spontaneous combustion is not uncommon, but it takes time for heat to build. "In industrial processes, the storage or disposal of oily rags in piles can allow them to self heat, or the combustion process could have been accelerated due to heat created from industrial processes like equipment operation or friction," reads the report. "Piles of straw, coal and even large manure piles can spontaneously combust."

Suppose a family member or friend had read this article and then asked you to explain what is meant by spontaneous combustion and how a pile of coal could suddenly burst into flames. How would you respond?

Note that the preceding newspaper passage itself provides a good hint in the first sentence. A review of Section 9.1 (especially Figure 9.1) would also be of some assistance. Additionally, you might want to consider Figure 9.5 and the corresponding explanation given in the original source paper by Park et al. [11]. Although strictly applicable to the case of an external heat source located

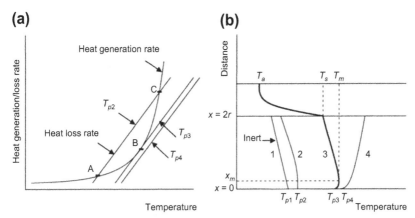

FIGURE 9.5   Dust layer heating and ignition: (a) concept of thermal runaway; (b) temperature distributions in an asymmetrically heated layer [11]. Symbols are defined as follows: T_p (hot-plate temperature); A, B, and C (points of intersection of rate of heat loss line with rate of heat generation curve); T_a (ambient temperature); T_m (maximum layer temperature); T_s (top-surface temperature); $x = 0$ (lower boundary of layer in contact with heated surface); and $x = 2r$ (upper boundary of layer).

beneath a dust layer, the overall concept is fundamentally the same—i.e., ignition occurring when the rate of heat generation exceeds the rate of heat loss. This is somewhat analogous to the relationship between suppressant requirement and delivery shown earlier in Figure 7.2.

Given that the issue of dust layer/pile combustion exists in industry, how can such hazards be identified and the ensuing risk assessed? One approach is through the use of graphical techniques such as fault tree analysis (FTA) and event tree analysis (ETA). FTA is a deductive method in which specific factors (e.g., fuel, oxidant, and ignition source) leading to a general undesired incident (e.g., fire) are identified. ETA, on the other hand, is an inductive method in which an event (e.g., fire) is further examined to determine eventual outcomes based on the success or failure of various safety measures or barriers (e.g., automatic extinguishing). Fault trees are therefore usually associated with prevention efforts and event trees with mitigation (or protection) efforts.

Figure 9.6 demonstrates the application of FTA and ETA to a case study of self-heating and self-ignition of coal in underground storage silos at a power plant in Finland [4]. Representation of a fault tree and an event tree together in this manner is called a bow-tie diagram, with the overall technique being termed bow-tie analysis (BTA). BTA is a powerful tool for hazard identification and risk assessment and is at the forefront of current engineering tools in the field of process safety.

Compare Figure 9.6 with the generic bow-tie diagram shown in Figure 9.7. Is the bow-tie shape apparent in each of Figures 9.6 and 9.7? Can you see possible uses for this technique in your own hazard identification and risk assessment efforts, both in general and specifically in relation to dust fires and dust

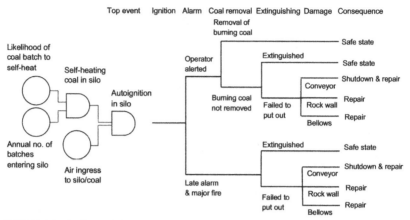

FIGURE 9.6 Fault tree (left) and event tree (right) for identifying hazards and assessing the risk of self-heating and self-ignition of coal stored in underground silos [4].

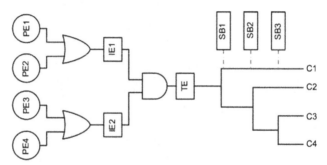

FIGURE 9.7 Generic bow-tie with fault tree on left of TE (top event) and event tree on right [23]. Other symbols are defined as follows: PE (primary event), IE (intermediate event), SB (safety barrier), and C (consequence).

explosions? Think of the possibilities if the bow-tie included prevention barriers in the fault tree portion, as well as the already present mitigation barriers in the event tree.

For those readers not familiar with FTA and ETA, the basic logic gates in Figure 9.7 are explained as follows:

And Gate—meaning that all inputs are required to generate the output:

Or Gate—meaning that only one input is required to generate the output:

(Note that the fault tree in Figure 9.6 shows only *and* gates.)

The paper by Khakzad et al. [23] (from which Figure 9.7 is drawn) adopts the 2008 Imperial Sugar Company explosion and fire [24] as a case study for bow-tie analysis. In their work we see the use of an *or* gate to determine that the required particulate fuel could have arisen from either sugar that was already airborne or sugar deposits on the floor. (Recall Figure 4.2.)

Can you now construct a generic fault tree for a dust layer fire and also one for a dust explosion? Think back to the discussion in Chapter 1 on the fire triangle (three requirements) and the explosion pentagon (five requirements). In either case, an *and* gate would be needed when considering the prerequisites that are all needed for the top event (fire or explosion). The use of *or* gates will, however, be needed to consider all possibilities leading to each of these fundamental underpinning factors. For example, what are the possible ignition sources for a dust fire? For a dust explosion? (See Section 8.1 in addition to the current chapter.)

REFERENCES

[1] Reddy PD, Amyotte PR, Pegg MJ. Effect of inerts on layer ignition temperatures of coal dust. Combustion and Flame 1998;114:41–53.

[2] Barton J, editor. Dust explosion prevention and protection. A practical guide. Rugby, UK: Institution of Chemical Engineers; 2002.

[3] Eckhoff RK. Dust explosions in the process industries, 3rd ed. Boston, MA: Gulf Professional Publishing/Elsevier; 2003.

[4] Sipila J, Auerkari P, Heikkila A-M, Tuominen R, Vela I, Itkonen J, Rinne M, Aaltonen K. Risk and mitigation of self-heating and spontaneous combustion in underground coal storage. Journal of Loss Prevention in the Process Industries 2012;25:617–22.

[5] Beamish BB, Barakat MA, St. George JD. Spontaneous-combustion propensity of New Zealand coals under adiabatic conditions. International Journal of Coal Geology 2001;45:217–24.

[6] Lebecki K, Dyduch Z, Fibich A, Sliz J. Ignition of a dust layer by a constant heat flux. Journal of Loss Prevention in the Process Industries 2003;16:243–8.

[7] Dyduch Z, Majcher B. Ignition of a dust layer by a constant heat-flux transport in the layer. Journal of Loss Prevention in the Process Industries 2006;19:233–7.

[8] Querol E, Torrent JG, Bennett D, Gummer J, Fritze J-P. Ignition tests for electrical and mechanical equipment subjected to hot surfaces. Journal of Loss Prevention in the Process Industries 2006;19:639–44.

[9] Janes A, Carson D, Accorsi A, Chaineaux J, Tribouilloy B, Morainvillers D. Correlation between self-ignition of a dust layer on a hot surface and in baskets in an oven. Journal of Hazardous Materials 2008;159:528–35.

[10] Gang L, Chunmiao Y, Peihong Z, Baozhi C. Experiment-based fire and explosion risk analysis for powdered magnesium production methods. Journal of Loss Prevention in the Process Industries 2008;21:461–5.

[11] Park H, Rangwala AS, Dembsey NA. A means to estimate thermal and kinetic parameters of coal dust layer from hot surface ignition tests. Journal of Hazardous Materials 2009;168:145–55.

[12] Joshi KA, Raghavan V, Rangwala AS. An experimental study of coal dust ignition in wedge shaped hot plate configurations. Combustion and Flame 2012;159:376–84.

[13] Krause U, Schmidt M, Lohrer C. A numerical model to simulate smouldering fires in bulk materials and deposits. Journal of Loss Prevention in the Process Industries 2006;19:218–26.

[14] Sweis FK. The effect of admixed material on the flaming and smouldering combustion of dust layers. Journal of Loss Prevention in the Process Industries 2004;17:505–8.

[15] Bideau D, Dufaud O, Le Guyadec F, Perrin L, Genin X, Corriou J-P, Caleyron A. Self-ignition of layers of powder mixtures: effect of solid inertants. Powder Technology 2011;209:81–91.

[16] Park H. Hot surface ignition temperature of dust layers with and without combustible additives. Master of Science Thesis. Worcester, MA: Worcester Polytechnic Institute; 2006.

[17] CSB. Investigation report—combustible dust hazard study. Report No. 2006-H-1. Washington, DC: U.S. Chemical Safety and Hazard Investigation Board; 2006.

[18] Astad J. Better identification of fire hazards needed. Occupational Health & Safety 2011;80:28–31.

[19] Hopkins A. Safety, culture and risk. The organizational causes of disasters. Sydney, Australia: CCH Australia Limited; 2005.

[20] Hopkins A. Failure to learn. The BP Texas City Refinery disaster. Sydney, Australia: CCH Australia Limited; 2009.

[21] Hopkins A, editor. Learning from high reliability organizations. Sydney, Australia: CCH Australia Limited; 2009.

[22] Lambert S. Fire caused by spontaneous combustion. Halifax, NS, Canada: The Chronicle Herald; October 13, 2012. Available at: thechronicleherald.ca/canada/147651-fire-caused-by-spontaneous-combustion; (last accessed March 24, 2013).

[23] Khakzad N, Khan F, Amyotte P. Dynamic risk analysis using bow-tie approach. Reliability Engineering and System Safety 2012;104:36–44.

[24] CSB. Investigation report—sugar dust explosion and fire—Imperial Sugar Company. Report No. 2008-05-I-GA Washington, DC: U.S. Chemical Safety and Hazard Investigation Board; 2009.

Chapter 10

Myth No. 9 (Oxidant): Oxygen Removal Must Be Complete to Be Effective

In this, our first specific look at the oxidant component of the explosion pentagon, we explore the notion that oxygen—by far the most common industrial oxidant—must be eliminated to correspondingly eliminate the dust explosion problem. The discussion here is closely related to that in the next chapter on the myth of creating a *safe* environment by means of oxygen removal.

Human beings do not actually require 21 volume % oxygen to survive. As we intuitively know, however, humans cannot exist in an atmosphere that is completely deprived of oxygen. Table 10.1 provides details of the effects on the human body of oxygen concentrations between 0 and 21 volume %, and illustrates that non-zero levels can be fatal.

Just as oxygen is essential for life, oxygen is essential for the combustion of industrial dusts. (We do need to acknowledge that other materials such as bromine, chlorine, and fluorine can also act as oxidants.) Here also, 21 volume % oxygen in the oxidizing atmosphere is not required for dust ignition and subsequent flame propagation. And again, non-zero oxygen percentages can interrupt the chemical (rather than physiological) processes necessary for a dust explosion to occur. Thus, oxygen does not have to be completely removed to effectively prevent dust explosions.

TABLE 10.1 Oxygen Deficiency Effects on the Human Body [1]

Oxygen percentage by volume	Physiological symptoms
23.5	Maximum safe level
21	Typical oxygen concentration in air
19.5	Minimum safe level
15–19	First signs of hypoxia; decreased ability to work strenuously; may induce symptoms in persons with heart, lung, or circulatory problems
12–15	Respiration increases with exertion; pulse up; impaired muscular coordination, perception, and judgment
10–12	Respiration further increases in rate and depth; poor judgement; blue lips
8–10	Mental failure; fainting; unconsciousness; ashen face; blue lips; nausea; vomiting; inability to move freely
6–8	Six minutes—50% probability of death; eight minutes—100% probability of death
< 6	Coma in 40 seconds, followed by convulsions; respiration ceases; death

(Original source is BP [2])

When complete inert gas blanketing of an enclosed process unit is undertaken, it is likely that the reasons are related to practical considerations such as ensuring the integrity of the prevention scheme itself. As noted by Barton [3], dust explosions can indeed be eliminated if air or other reactive atmospheres (e.g., pure oxygen or some other oxidant) are completely excluded by means of inert gas substitution. The existence of the explosibility parameter termed *limiting oxygen concentration,* however, means that oxygen reduction to a sufficiently low level can also be used to achieve the same goal [3].

10.1 LIMITING OXYGEN CONCENTRATION

Table 5.1 indicates that the limiting oxygen concentration in the atmosphere for flame propagation in a dust cloud is given the abbreviation LOC. As we will see shortly, the inert gas giving rise to a particular LOC must also be specified; LOC is expressed in percentage on a volume basis (equivalent to a mole basis for gases at conditions of ideal behavior).

By definition, the LOC is the highest oxygen concentration in a dust/air/inert gas mixture at which an explosion fails to occur [3]. To the end of this definition, Hoppe and Jaeger [4] add the caveat *under defined experimental conditions*; the reason for this addition is discussed at the end of the current section.

Typical LOC values are given in Table 10.2 for the inerting of various dust types with nitrogen. To account for operating errors and upset conditions, a safety factor is normally applied to the measured LOC (e.g., the LOC percentage minus a further 2 volume %) [4].

Consistent with the note at the end of Section 8.2, details such as particle size distribution are not given in Table 10.2 because the focus is on providing

TABLE 10.2 Examples of Limiting Oxygen Concentration Values for Several Representative Dust Types [4]

Dust type	LOC (with nitrogen) [volume %]
Pea flour	15.5
Organic pigment	12.0
Cadmium stearate	12.0
Wheat flour	11.0
High-density polyethylene	10.0
Sulfur	7.0
Aluminum	5.0

general examples that are broadly representative of a range of materials. It is reasonable to conclude from Table 10.2 that nitrogen inerting requirements for highly reactive metals such as aluminum are more stringent than for organic dusts. The data in Table A2 in the appendix in Eckhoff [5] support this conclusion and give particle sizes for the dusts listed there in terms of the mass median diameter. (Table A2 in Eckhoff [5] carries the heading of "Maximum permissible oxygen concentration for inerting dust clouds in atmospheres of oxygen and nitrogen", meaning that these are LOC data.)

The concept of a limiting oxygen concentration also applies to hybrid mixtures of a combustible dust and a flammable gas. (See Section 5.4.) Figure 10.1 shows the results of nitrogen inerting tests conducted for the tungsten/hydrogen system in a 20-L chamber with electric spark ignition; a more energetic ignition source was not required because of the presence of hydrogen [6].

A predetermined value of the maximum explosion pressure [e.g., 1 bar(g)] is usually used as the explosibility criterion for dust flammability limit parameters such as LOC and MEC (minimum explosible concentration). Denkevits [6] argues that for systems in which a detonation could arise, the maximum rate of pressure rise is a better measure of explosibility. When this logic was applied to hybrid mixtures containing hydrogen and extrapolated to zero rate of pressure rise in Figure 10.1, limiting oxygen concentration estimates in the range of 6–9 volume % were obtained for the various experimental conditions [6]. (These are the y-intercept values appearing as the last term in each straight-line equation in the legend to Figure 10.1.)

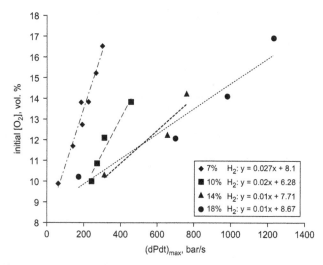

FIGURE 10.1 Measurement of limiting oxygen concentration with nitrogen for hybrid mixtures of tungsten (0.2–0.55 μm particle size and 2000 g/m^3 dust concentration) and hydrogen (7, 10, 14, and 18 volume %) [6].

Other factors in addition to the co-presence of a flammable gas can influence the determination of limiting oxygen concentrations via closed-vessel explosibility testing. Figure 10.2 illustrates the role of vessel volume and ignition energy in this regard for Pittsburgh seam bituminous coal dust. (See Section 6.1.) Here, we see that for this material, ignition energies of 2.5 kJ (plot c) and 5 kJ (plot d) in a 20-L chamber yield lower oxygen percentages for inerting with nitrogen than an ignition energy of 10 kJ in a 1-m^3 chamber (plot a). The 20-L ignition energy had to be lowered to 1 kJ (plot b) in these tests to give results comparable those for the 1-m^3 vessel.

By way of background, a 10-kJ ignition energy in a 1-m^3 explosion chamber is the general benchmark for measurement of most explosibility parameters

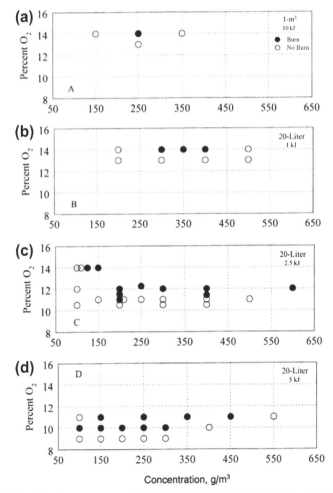

FIGURE 10.2   Limiting oxygen concentration data (nitrogen inerting) for Pittsburgh seam bituminous coal dust showing the influence of vessel volume and ignitor energy [7].

(P_{max}, K_{St}, MEC, and LOC). Testing at the laboratory scale of 20 L requires care in ensuring an appropriate ignition energy is used so as not to *overdrive* the system; 10 kJ is still suitable (and, in fact, required) for P_{max} and K_{St} determination, but lower ignition energies must be used for MEC and LOC. (The occurrence of overdriving and its implications for dust explosibility testing are covered in some detail in Chapter 19.)

The practical implication of Figure 10.2 is the familiar need for standardized equipment and standardized procedures when testing for LOC and all other explosibility parameters. Testing protocols developed by recognized standards organizations on the basis of best-practice methodologies are an absolute requirement, as is the testing of actual representative industrial dust samples. These points have already been made several times in this book; it is critical that end users of dust explosibility data recognize their importance.

10.2 CANDIDATE INERT GASES

Barton [3] provides a listing of the typical inert gases used in industry for dust explosion prevention purposes, along with general pros and cons for each. Available candidates include carbon dioxide, nitrogen, flue gases, argon, helium, and steam [3]. Common issues of concern are (i) effectiveness, (ii) availability, (iii) cost, (iv) compatibility with process materials and products, and (v) introduction of new hazards [3]. The last of these points is addressed at length in the next chapter.

Hoppe and Jaeger [4] provide an explanation for the enhanced effectiveness of some inert gases over others. They state that multiatomic molecules have a greater capacity for energy absorption because of their larger number of free molecular oscillation degrees; hence, triatomic CO_2 and diatomic N_2 would be expected to be more effective than monatomic Ar [4].

These considerations are consistent with the reasoning of Li et al. [8], who used the physical property of specific heat capacity as a basis for explaining the influence of different inert gases on dust explosibility. As shown in Table 10.3, the order from highest to lowest specific heat capacity is $CO_2 \rightarrow N_2 \rightarrow Ar$.

While the preceding generalizations are, of course, helpful as a guide, they are not a substitute for actual LOC test data. (Recall the final paragraph of the

TABLE 10.3 Specific Heat Capacity of Candidate Inerting Gases at 273 K [8]

Gas	Specific heat capacity [J/mol·K]
Argon	20.72
Nitrogen	28.78
Carbon dioxide	37.23

previous section.) For a specific example, we turn again to the work of Li et al. [8], who conducted LOC tests for magnesium in a 20-L chamber with a 2-kJ ignition energy over a dust concentration range of 50–450 g/m^3. The upper end of this range represents the stoichiometric concentration for the magnesium sample tested (mass median diameter of 47 μm) [8]. The LOC results they obtained are shown in Table 10.4.

Given the previous discussion in this section, it is not unexpected to see the following sentence in the paper by Li et al. [8] (p. 183): "Surprisingly, nitrogen as inerting agent yielded the highest LOC." In other words, nitrogen was the most effective inert gas for the scenario investigated. It is clear the authors were also aware of possible metal dust/inert gas reaction because of their comment that both nitrogen and carbon dioxide can react with magnesium [8]. They surmise that the energy liberated during any initial reaction between magnesium and available oxygen may have been too low to facilitate magnesium/nitrogen or magnesium/carbon dioxide reactions [8].

10.3 REALITY

The use of an inert gas to render an oxidizing atmosphere incapable of supporting combustion can be an effective means of dust explosion prevention. Complete removal of oxygen by substitution of air with an inert gas—known as *total inerting*—is not often undertaken in industrial practice [4].

The more common approach in the process industries is to dilute the oxidizing atmosphere by introduction of an inert gas [4]. In this manner, the oxygen level is reduced to a volume percentage at which an explosion cannot occur even in the presence of a normally adequate ignition source; the highest oxygen concentration for such non-explosibility is the limiting oxygen concentration, or LOC.

Table 10.5 illustrates the existence of a limiting oxygen concentration even for a reactive powder having—in relative terms for dusts—high values of P_{max} and K_{St} and a low value of MIE. This particular material, nano-size aluminum or Alex, finds potential application as an additive to explosives to increase detonation velocity and detonation pressure [9].

TABLE 10.4 Limiting Oxygen Concentration Values for Magnesium with Various Inerting Gases [8]

Gas	LOC [volume %]
Argon	4.0
Nitrogen	6.8
Carbon dioxide	5.5

TABLE 10.5 Alex (Nano-Size Aluminum) Dust Explosibility Parameters [9]

P_{max} [bar(g)]	K_{St} [bar m/s]	MIE (with inductance) [mJ]	LOC (with nitrogen) [volume %]
9.4	322	1–3	5

10.4 WHAT DO *YOU* THINK?

Hoppe and Jaeger [4] use the term *partial inerting* to distinguish between total inerting as described in the previous section and dilution of the oxidizing atmosphere to a set margin below the LOC. Partial inerting is used by Eckhoff [10] to also mean dilution of the oxidizing atmosphere with an inert gas, but not to the extent that the LOC is attained. Partial inerting in this context refers to the type of behavior displayed in Figure 10.3 for the explosibility of brown coal dust at reduced oxygen levels.

Here, we see it is not until 11 volume % oxygen that the limiting oxygen concentration is reached and explosions are no longer possible. There is, however, a general lowering of overpressure and rate of pressure rise at oxygen concentrations of 14, 12, and 11.5 volume %. Eckhoff [10] argues that this reduction in consequence severity affords practitioners an *additional degree of freedom* in their explosion prevention and mitigation efforts. To illustrate this point, he gives the examples of potentially smaller vent area requirements and increased effectiveness of automatic suppression devices [10].

Eckhoff [10] cautions though that with suppression, details on system design and performance are required to quantify the effects of partial inerting. This note of concern seems warranted based on considerations covered earlier in the current book. Is there a parallel to be drawn between gaseous and solid inertants when they are used in less than the full amount needed for explosion prevention? (You may wish to review the material in Chapter 7.)

Figure 10.4 from the work of Li et al. [8] (magnesium dust testing in a 20-L chamber with a 2-kJ ignition energy) shows the occurrence of a previously defined phenomenon—this time for carbon dioxide inerting at 15.3 volume % oxygen (i.e., at an oxygen concentration significantly higher than the LOC value of 5.5 volume % given in Table 10.4). Looking at the carbon dioxide data for magnesium dust concentrations greater than 1500 g/m³, can you identify the phenomenon referred to in the previous sentence? (See Section 7.3 and Figure 7.5.) Would you agree that the parameter in question is material (inert gas)-specific?

To conclude, we return to Figure 10.3 in which Eckhoff [10] notes the occurrence of an upper flammable limit (maximum explosible concentration) at reduced oxygen concentrations. Do you see an upper limit for brown coal

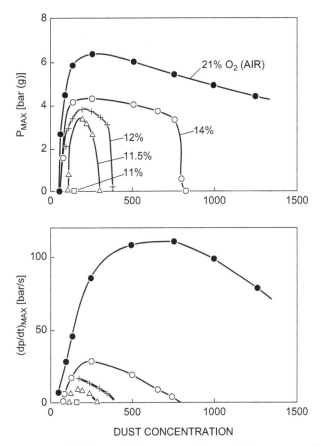

FIGURE 10.3 Inerting effectiveness of nitrogen on maximum explosion pressure and maximum rate of pressure rise of brown coal. Tests were conducted in a 1-m³ chamber at an initial temperature of 150°C [10]. Original source is Wiemann [11].

at about 800 g/m³ dust concentration and 14 volume % oxygen? How does this 14% oxygen curve compare with the 21% oxygen curve in Figure 10.3? How does it compare with the polyethylene and coal dusts curves in Figure 5.4 (Section 5.5), which would have been determined in a 21 volume % oxygen environment?

Dilution with an inert gas can therefore be expected to have a beneficial effect from a prevention perspective on explosibility parameters other than overpressure and rate of pressure rise. The impact of oxygen level reduction on the MEC of aluminum and magnesium samples is shown in Figure 10.5. Would you expect a similar increase in MIE at oxygen concentrations lower than the usual 21 volume %? (You can trust your intuition on this question or source the original paper by Eckhoff [10] for confirmation.)

An Introduction to Dust Explosions

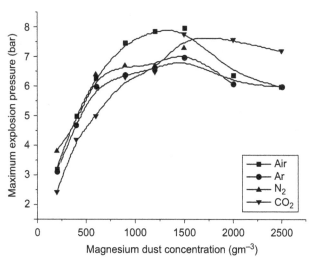

FIGURE 10.4   Inerting effectiveness of argon, nitrogen, and carbon dioxide on explosions of magnesium (mass median diameter of 47 μm) at 15.3 volume % oxygen [8].

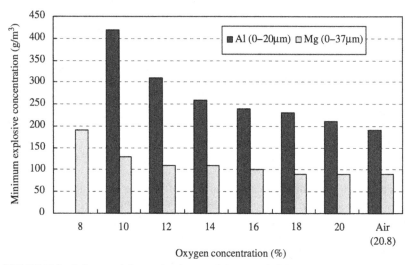

FIGURE 10.5   Influence of nitrogen inerting on minimum explosible concentration of aluminum and magnesium dust explosions in a modified Hartmann apparatus with spark ignition [12].

REFERENCES

[1] CSB. Case study. Confined space entry—worker and would-be rescuer asphyxiated. Valero Energy Corporation Refinery. Report No. 2006-02-I-DE. Washington, DC: U.S. Chemical Safety and Hazard Investigation Board; 2006.

[2] BP Process Safety Series. Hazards of nitrogen and catalysts handling. Rugby, UK: Institution of Chemical Engineers; 2004.

[3] Barton J, editor. Dust explosion prevention and protection. A practical guide. Rugby, UK: Institution of Chemical Engineers; 2002.

[4] Hoppe T, Jaeger N. Reliable and effective inerting methods to prevent explosions. Process Safety Progress 2005;24:266–72.

[5] Eckhoff RK. Dust explosions in the process industries, 3rd ed. Boston, MA: Gulf Professional Publishing/Elsevier; 2003.

[6] Denkevits A. Hydrogen/dust explosion hazard in ITER: effect of nitrogen dilution on explosion behaviour of hydrogen/tungsten dust/air mixtures. Fusion Engineering and Design 2010;85:1059–63.

[7] Going JE, Chatrathi K, Cashdollar KL. Flammability limit measurements for dusts in 20-L and 1-m^3 vessels. Journal of Loss Prevention in the Process Industries 2000;13:209–19.

[8] Li G, Yuan CM, Fu Y, Zhong YP, Chen BZ. Inerting of magnesium dust cloud with Ar, N_2 and CO_2. Journal of Hazardous Materials 2009;170:180–3.

[9] Kwok QSM, Fouchard RC, Turcotte A-M, Lightfoot PD, Bowes R, Jones DEG. Characterization of aluminum nanopowder compositions. Propellants, Explosives, Pyrotechnics 2002;27:229–40.

[10] Eckhoff RK. Partial inerting—an additional degree of freedom in dust explosion protection. Journal of Loss Prevention in the Process Industries 2004;17:187–93.

[11] Wiemann W. Influence of temperature and pressure on the explosion characteristics of dust/air and dust/air/inert gas mixtures. In: Cashdollar KL, Hertzberg M, editors. Industrial Dust Explosions. ASTM Special Technical Publication 958. Philadelphia, PA: American Society for Testing and Materials; 1987.

[12] Nifuku M, Koyanaka S, Ohya H, Barre C, Hatori M, Fujiwara S, Horiguchi S, Sochet I. Ignitability characteristics of aluminum and magnesium dusts that are generated during the shredding of post-consumer wastes. Journal of Loss Prevention in the Process Industries 2007;20:322–9.

Chapter 11

Myth No. 10 (Oxidant): Taking Away the Oxygen Makes Things Safe

In this second chapter dealing with the oxidant component of the explosion pentagon, our focus shifts to the belief that an industrial process can be made *safe*—i.e., placed in a state in which there are no hazards, with subsequent zero risk and an absolute guarantee that nothing will go wrong. We examine the specific issue of oxygen removal as detailed in the preceding chapter, as well as a few more general process safety considerations. The current chapter addresses the topic at hand from two perspectives: (i) what can happen when safety devices fail and procedural measures are inadequate, and (ii) how unanticipated new hazards can be created if process changes are not well managed.

Is it safe? I was once asked this question by the manager of a facility that had experienced a dust explosion. No one was hurt in the incident, but there was significant equipment damage and a resulting period of business interruption. The consultant with whom I was working had made several recommendations for additional prevention and mitigation measures according to applicable codes and standards, and these measures had been implemented.

How then does one answer this question posed by a concerned, well-meaning business manager? As my consultant friend has cautioned me more than once, you don't respond with a yes or no answer. Because nothing is safe and to imply it is means accepting enormous personal and professional liability; yet at the same time, the process needs to be restarted as soon as possible to avoid continued loss of revenue. So it really comes down to a matter of risk.

11.1 NOTHING IS SAFE

If it is neither wise nor feasible to address the issue of whether a process is safe, then perhaps a more appropriate question is whether a resumption of operations following an incident can, in fact, be recommended. The answer to this question (yes or no) depends on whether the residual risk is assessed to be acceptable based on relevant criteria (company, societal, regulatory, etc.).

An example in this regard is given in Figure 11.1, which shows the ALARP approach based on the principle of managing risk at a level as low as reasonably practicable [1]. (Note that the *P* in *ALARP* stands for *practicable*, not *possible*.) Vinem [2] explains that use of the ALARP principle requires examination of both risk levels and risk mitigation costs, with risk reduction measures being implemented according to cost effectiveness considerations.

Thus, a process can be made safer than it was before the introduction of additional safety measures, but it cannot be made 100% safe. More to the point, the potential benefits of attempting to drive process risk to zero will likely be grossly disproportionate to the cost involved. We should talk about safer alternatives, not safe processes.

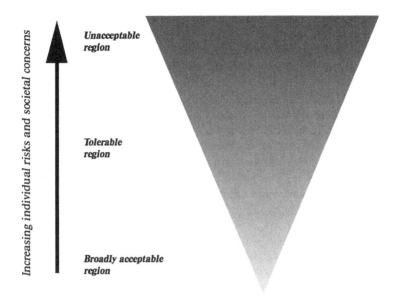

FIGURE 11.1   Representation of the ALARP (As Low As Reasonably Practicable) principle [1].

As an example, in Chapter 15 we will cover in some detail the concept of inherently safer design (with emphasis on the word *safer*). Inherently safer design, or ISD, is effective in matters of risk reduction because it involves the elimination of hazards at their source. Next in order of effectiveness are engineered (add-on) safety devices and systems, followed by procedural measures such as safe work procedures.

> Note: At this point, I should apologize for the inevitable use of the word *safe* in this book. It will undoubtedly appear here and there (as it just did)—mostly as a matter of grammatical form or because of industry convention, but not to imply absolute freedom from risk.

A key point to recognize about safety devices is that they might fail when called upon to act. Further, procedures might be ignored or not followed in spite of training in their use. Even the measures relied upon to create a safer work environment can contribute to an increase in risk and ultimately an undesired event.

As a case in point, we consider a dust explosion that occurred at a magnesium powder mixing plant. (All details given here are from REMBE.® [3]) A mixer was being gravity-filled from a container through a charge chute, with explosion prevention provided by an inert gas blanketing system. Powder flow to the mixer was interrupted, and the operator attempted to clear the plugged charge chute by prodding through an inspection hatch with a metal rod (rather than the recommended wooden rod). Frictional sparking ignited a dust cloud that formed when the blockage was cleared. The operator was severely burned by flames exiting the inspection hatch.

Counting the explosion pentagon elements in the preceding description, we see there are four: (i) fuel—magnesium powder, (ii) ignition source—friction spark, (iii) mixing—sudden flow of powder on release of the plug, and (iv) confinement—container, charge chute, and mixer. But what of the fifth component—oxidant? It turns out that the design of the inert gas system was flawed; air was able to enter the charge chute and the top of the mixer through the inspection hatch. Among other recommendations arising from the ensuing incident investigation (e.g., use of a vibratory feeder and a rod composed of non-sparking material), a key measure identified was maintaining the integrity of the inert gas supply by means of a smaller inspection hatch and a slightly pressurized system [3].

Vigilance in guarding against loss of inerting is essential when utilizing oxygen removal as a means of explosion prevention. Figure 11.2 demonstrates this point by showing the increase in oxygen concentration in a nitrogen-inerted reactor after opening of a manhole and addition of powder. After about 3 minutes, the oxygen percentage rises to a level above likely LOC values; after 7–10 minutes the percentage is essentially that at ambient conditions. Even with a significant increase in nitrogen flow rate after closing the manhole, pre-opening oxygen concentrations are not reached until almost 1 hour after the initial upset. Glor [4] explains these observations by diffusion of the reactor inert atmosphere into the external environment as well as air-entrainment by the powder during addition to the reactor. He also lists various countermeasures to prevent loss of inerting [4].

The REMBE® booklet [3] describes a second incident involving loss of inert gas (intentional in this case). An acrylonitrile-butadiene-styrene (ABS) flash dryer had been shut down for cleaning. In contravention of the established

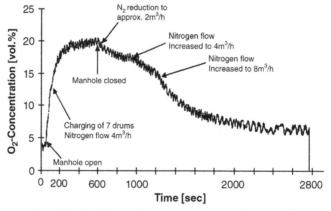

FIGURE 11.2   Loss of inerting (increase in oxygen concentration) in a reactor following manhole opening and powder addition [4].

procedure, the inert gas (nitrogen) flow had been immediately switched to air flow and the inspection hatch opened. Consider the explosion pentagon elements apparent at this point: (i) fuel—ABS dust, (ii) oxidant—oxygen in air, and (iii) confinement—flash dryer.

Because the dryer had not been sufficiently cooled, ABS dust deposits began to self-heat and a fire resulted. (Recall the discussion of layer fires in Chapter 9.) Operators used a manual fire extinguisher in an attempt to address this event, but the extinguishing powder had the unintended effect of creating an airborne cloud of ABS dust that was ignited by the original fire. With the addition of mixing and an ignition source, the pentagon was complete and a dust explosion occurred. Both operators were injured and the flash dryer was damaged [3].

11.2 INTRODUCTION OF NEW HAZARDS

As mentioned in Section 10.2, the possible introduction of new hazards is a concern common to all inert gases employed for explosion prevention. So while one is attempting to remove one set of hazards related to dust explosions, vigilance is also required to ensure that other hazards arising from the use of an inert gas are identified and adequately managed.

This requirement is so critical in the process industries that technical documents on inert gas blanketing usually make some reference to the possible creation of new hazards and the additional safety measures required. Examples include

- Reaction of carbon dioxide with aluminum dust [5,6]
- Reaction of nitrogen with magnesium dust at high temperatures [5,6]
- Asphyxiation hazard due to inert gas (requiring detectors, alarms, etc.) [5,7–9]
- Leakage of inert gas from systems under pressure [3,8]
- Electrostatic discharges when carbon dioxide is drawn from high-pressure or cryogenic tanks [8]
- Ignition due to the pump used for system evacuation during vacuum inerting [8]
- Ignition in the absence of oxygen when processing pyrotechnics or explosives in which each fuel molecule has its own oxidizing group [10].

The introduction of a new hazard through a measure designed to reduce risk from a different hazard is certainly not unique to inert gas blanketing. There are many other process industry examples where this can occur, such as the use of "gas blows" to remove debris and clean out natural gas pipelines.

Figure 11.3 shows a gas blow at the Kleen Energy site in Middletown, CT, one week before a series of these undertakings led to an explosion that killed six workers [11]. The fuel for the explosion was the natural gas being forced through a pipe at a pressure of 4.5 MPa (650 psi) and eventually exiting into a congested, outdoor work area where a number of potential ignition sources existed [11]. (All five elements of the explosion pentagon can be seen in the preceding sentence; this is a special case of *confinement*, as treated in Chapter 17.)

FIGURE 11.3   Gas blow used to clean fuel gas piping at the Kleen Energy site in Middletown, CT, one week before the incident on February 7, 2010. The darker plume in the center of the picture is indicative of debris being blown from the line [11].

You might be surprised to learn that natural gas was the propelling medium of choice in this incident. Its use was obviously not to remove oxygen to address a combustion hazard as in the case of inert gas blanketing; the exact opposite occurred with the introduction of a flammable gas to the operation of pipeline cleaning. However, in the set of recommendations drawn from its investigation [11], the U.S. Chemical Safety Board (CSB) noted that there are several inherently safer alternatives to using natural gas for the purpose of cleaning fuel pipes. (As previously mentioned, Chapter 15 deals with the topic of inherently safer design.) These alternatives include (i) pigging with air or nitrogen; (ii) air, nitrogen, or steam blows; and (iii) cleaning with water or chemicals [11].

The CSB recommendations [11] further note that the preceding techniques, while aimed at the removal of fire and explosion hazards, may also introduce new hazards such as nitrogen asphyxiation. One would perhaps note as well that even the use of a seemingly benign substance like air would entail propelling a gas through a pipeline under significant pressure.

On the matter of oxygen-deficient atmospheres (see Table 10.1), attention is drawn to an excellent training vehicle produced by the CSB—a video on the hazards of nitrogen asphyxiation [12]. Watching this video would represent 12 minutes very well spent. It provides both general process safety lessons and specific information on the events surrounding the deaths of two contract workers performing maintenance activities near the opening to the nitrogen-inerted reactor shown in Figure 11.4 [13]. The video [12] and accompanying case study [13] demonstrate the insidious nature of confined-space entry incidents in which would-be rescuers often succumb to the effects of low oxygen concentrations.

FIGURE 11.4 Unsecured confined space access opening to a reactor at the Valero Energy Corporation Refinery in Delaware City, DE [13].

11.3 MANAGEMENT OF CHANGE

Vaughen and Klein [14] (p. 414) comment that "[c]hanges to hazardous processes—emergency, simple and complex—must be managed." Hence arises the term *management of change* or *MOC*.

Management of change is a critical element of any safety management system, particularly those designed for process safety management such as the 20-element, risk-based approach [15] recently developed by the Center for Chemical Process Safety (CCPS) in the United States. (Safety management systems are also discussed briefly in Chapter 21.)

In the process industries, the most well-known example of the need for MOC—at least with respect to temporary modifications—is the 1974 Flixborough, UK, disaster in which a large release of cyclohexane exploded, killing 28 people and causing significant facility damage [16]. With respect to the subject of the current chapter, MOC is essential to manage the risk of changes to explosion prevention and mitigation systems—changes that may unintentionally introduce new hazards as described in the previous section.

For example:

- What if a decision is made to substitute one inert gas for another—will this new gas react with the dust being handled?
- What if a different vacuum pump is used for a vacuum inerting system—has the suite of potential ignition sources been affected?
- What if a switch to a pressurized inerting system is made—will leakage into the work environment increase?
- What if the operating temperature envelope is altered—will the inert atmosphere be lost due to steam condensation?
- What if a new process vessel is brought into service—is the material of construction compatible with the inerting agent?

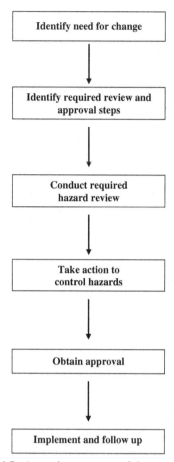

FIGURE 11.5   A generic management of change protocol [17].

- What if inert gas blanketing represents a completely new explosion prevention philosophy for a company—is the company prepared for prevention of asphyxiation hazards as well as explosion hazards?

To properly address these and other change-related issues, a formal, structured MOC process is indispensable. Figure 11.5 gives a generic framework in this regard. Embodied within this protocol, as in all process safety management elements, are the four essential functions of *plan, do, check,* and *act.*

11.4 REALITY

No industrial process is *safe* in the sense that it is absolutely free from risk. However, *safer* processing alternatives might exist in which the risk of fire or

FIGURE 11.6   Siwek 20-L apparatus for determination of dust explosibility—P_{max}, K_{St}, MEC, and LOC (manufactured by Kuhner AG, Basel, Switzerland) [18].

explosion is substantially reduced. It is also important to recognize that new, unexpected hazards can be introduced by the very act of removing an existing hazard. There are numerous warnings in the technical literature on the possible creation of an asphyxiation hazard when oxygen is replaced by nitrogen or other inert gases to address hazards related to dust explosions.

As a final example, reference is made to recent research on nano-size dust explosibility as determined in a Siwek 20-L chamber, a photograph of which is shown in Figure 11.6. (See also Section 8.2 and Chapter 19.) This represents a closed system because the apparatus is used to determine dust explosibility parameters based on overpressure and/or rate of pressure rise development. Dust (*fuel*) is placed in the 1-L storage container shown in the right foreground of Figure 11.6. This external reservoir is then pressurized with air (*oxidant*), and the dust is dispersed (*mixing*) through a pipe bend and valve into the spherical 20-L vessel (*confinement*) and ignited by one or more centrally located chemical ignitors of known energy (*ignition source*).

Wu et al. [19] used a 20-L apparatus to conduct a novel investigation of the effect of pneumatic transport velocity on the explosibility of 35-nm aluminum, 35-nm iron, and 30-nm titanium powder samples. They varied both the carrier gas itself (air or nitrogen) and its velocity (by manipulating the dispersion pressure); no external ignition source (i.e., chemical ignitor) was used. The titanium exploded in air at all velocities but not in nitrogen, iron exploded in air at all but the lowest velocity and not in nitrogen, and aluminum did not explode in either air or nitrogen [19].

FIGURE 11.7 Explosions of 30-nm titanium powder in the dust transport system of the Siwek 20-L chamber with air dispersion: (a) at the pipe bend after the external dust reservoir; (b) in the external dust reservoir; (c) again, in the external dust reservoir with the cover off post-explosion [19].

Figure 11.7 shows the physical evidence of the explosions that occurred in the dust transport system when conveying the 30-nm titanium powder with air. Given that the minimum ignition energy of each of the three metal dusts was less than 1 mJ, it is surmised that either frictional or electrostatic sparking was the source of ignition [19].

In essence, then, oxygen removal by replacement with nitrogen effectively prevented the nano-titanium powder from exploding during these experimental tests. This is not the end of the story, however.

In our own dust explosion research laboratory at Dalhousie University, we recently completed a similar investigation of micron- and nano-titanium explosibility using a 20-L apparatus [20]. (These are the samples described in Section 8.3 and Table 8.2.) We knew of the work of Wu et al. [19] and were able to replicate the scenario illustrated by Figure 11.7. A decision was thus

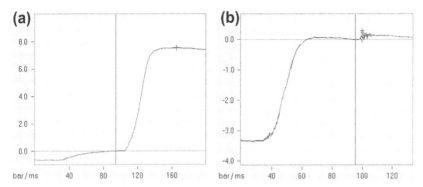

FIGURE 11.8   Pressure/time trace (Siwek 20-L chamber) for (a) −325 mesh (< 45 μm) titanium at 1750 g/m^3 with air dispersion; (b) 40–60 nm titanium at 100 g/m^3 with nitrogen dispersion [20].

made to disperse the nano-titanium from the external dust reservoir by means of nitrogen carrier gas into an initially elevated oxygen environment in the explosion chamber. Because of the delay time before ignition source activation, mixing of the pure nitrogen stream and oxygen-rich atmosphere would ensure the oxygen concentration in the 20-L chamber was 21 volume % at the time of ignition.

Figure 11.8 gives typical pressure/time traces from this testing; here, time appears on the x-axis and pressure on the y-axis. Trace (a) is for a micron-size sample with normal air dispersion. At time zero, the dispersion valve begins to open; and 35 ms later, air starts to enter the 20-L chamber, and the pressure rises to atmospheric. (The chamber is partially evacuated prior to initiating the test sequence.) After a 60-ms delay time has elapsed (indicated by the vertical line at 95 ms), the chemical ignitors are actuated and the dust explodes, causing the pressure to rise to a peak value followed by a gradual decrease in pressure as the vessel contents cool.

Trace (b) is for a nano-size sample with nitrogen dispersion. Here, we see an overlap of the dust dispersion and explosion steps, with the ignitors firing after the dust has exploded. While an explosion in the external dust reservoir and pipe bend was avoided by transport with inert nitrogen, frictional or electrostatic sparking ignited the titanium upon first contact with the oxygen molecules in the 20-L chamber. (The pressure values on the y-axis must be interpreted in light of the fact that the pressure transducers used are piezoelectric and hence measure only dynamic pressure changes, not static values.)

It is not too much of a stretch to imagine a similar industrial situation in which a highly reactive dust is being transported by an inert carrier gas. This could be justifiably considered as a *safer* approach to powder conveyance than using air as the carrier gas and attempting to eliminate ignition sources. But it is far from *safe* given the possibility of a sudden loss of inerting and encounter of the dust with oxygen. And, of course, there may well be *safer* yet means of bulk powder transport than the scheme just described.

11.5 WHAT DO *YOU* THINK?

Martin Glor [4] (p. 2) writes

According to the author's experience, most pharmaceutical companies are well aware of the explosion hazards and make appropriate efforts to minimize the explosion risk (explosion probability as well as explosion severity). However, experience also shows that management of changes is not always dealt with in a prospective way in the field of explosion prevention. The cumulative effect of small single changes of the process, operation, or product may lead to a substantial increase of the explosion hazard. In addition, increasing turnover of personnel may lead to a lack of knowledge, which can only be compensated with increased training.

Here are some questions to consider:

- Would you agree that the preceding comments—at least the ones related to management of change—could likely be extended to most industries in which combustible dusts are handled? Why or why not?
- Can you identify some specific examples of inadequate MOC in the dust explosion literature? (A good starting point would be the 2008 Imperial Sugar Company explosion [21], for which Taveau [22] cited the MOC issues related to increased confinement brought about by installation of a cover over a conveyor belt.)
- Can you identify some general examples of inadequate MOC in the general process safety literature? (The case study book by Trevor Kletz [16] is an excellent resource.)
- In the second paragraph of Section 6.5, I suggested a few questions related to process changes leading to finer-sized dust and requiring a rethinking of explosion risk reduction strategies. You might reconsider these points in light of the discussion in the current chapter and Glor's comments on changes in process, operation, or product [4].
- The last sentence in the preceding quote from Glor [4] speaks of the dual issues of loss of corporate memory and management of organizational change. How do you account for these factors within your organization?

REFERENCES

[1] HSE. Reducing risks, protecting people. HSE's decision-making process London, UK: Health and Safety Executive; 2001. (Contains public sector information published by the Health and Safety Executive and licensed under the Open Government Licence v1.0).

[2] Vinem E. Ethics and fundamental principles of risk acceptance criteria. Safety Science 2012;50:958–67.

[3] REMBE®. REMBE®'s booklet of safety and security (BOSS). Dust explosions. A comprehensive guideline to industrial explosion protection including scientific basics, case studies about incidents, prevention methods and constructive protection measures. Brilon, Germany: REMBE® GMBH · Safety + Control; 2012, Available at: http://www.rembe.us/case-studies/documents/GBA-BOSS-09064-2.pdf; last accessed January 15, 2013.

[4] Glor M. A synopsis of explosion hazards during the transfer of powders into flammable solvents and explosion preventative measures. Pharmaceutical Engineering 2010;30:1–8.

[5] Abbasi T, Abbasi SA. Dust explosions—cases, causes, consequences, and control. Journal of Hazardous Materials 2007;140:7–44.

[6] Li G, Yuan CM, Fu Y, Zhong YP, Chen BZ. Inerting of magnesium dust cloud with Ar, N_2 and CO_2. Journal of Hazardous Materials 2009;170:180–3.

[7] Ebadat V. How do you determine when an inert gas environment is the best option to reduce dust explosion risk (non-solvent)? Combustible Dust & Static Electricity Q&A. Chilworth, 2012. Available at: http://pbs.canon-experts.com/expert/dr-vahid-ebadat/; last accessed October 22, 2012.

[8] Hoppe T, Jaeger N. Reliable and effective inerting methods to prevent explosions. Process Safety Progress 2005;24:266–72.

[9] Sipila J, Auerkari P, Heikkila A-M, Tuominen R, Vela I, Itkonen J, Rinne M, Aaltonen K. Risk and mitigation of self-heating and spontaneous combustion in underground coal storage. Journal of Loss Prevention in the Process Industries 2012;25:617–22.

[10] Pegg MJ, Amyotte PR, Lightfoot PD, Lee MC. Dust explosibility characteristics of azide-based gas generants. Journal of Loss Prevention in the Process Industries 1997;10:101–11.

[11] CSB. Urgent recommendations from Kleen Energy investigation. Washington, DC: U.S. Chemical Safety and Hazard Investigation Board; 2010.

[12] CSB. Hazards of nitrogen asphyxiation (video). Washington, DC: U.S. Chemical Safety and Hazard Investigation Board; 2006. Available at: http://www.csb.gov/investigations/detail. aspx?SID=25&Type=2&pg=1&F_All=y; last accessed October 22, 2012.

[13] CSB. Case study. Confined space entry—worker and would-be rescuer asphyxiated. Valero Energy Corporation Refinery. Report No. 2006-02-I-DE. Washington, DC: U.S. Chemical Safety and Hazard Investigation Board; 2006.

[14] Vaughen BK, Klein JA. What you don't manage will leak: a tribute to Trevor Kletz. Process Safety and Environmental Protection 2012;90:411–18.

[15] CCPS (Center for Chemical Process Safety). Guidelines for risk based process safety. Hoboken, NJ: John Wiley & Sons, Inc; 2007.

[16] Kletz T. What went wrong? Case histories of process plant disasters and how they could have been avoided, 5th ed. Oxford, UK: Gulf Professional Publishing; 2009.

[17] Amyotte PR, Goraya AU, Hendershot DC, Khan FI. Incorporation of inherent safety principles in process safety management. Orlando, FL: Center for Chemical Process Safety, American Institute of Chemical Engineers, Proceedings of 21st Annual International Conference—Process Safety Challenges in a Global Economy; 2006 (April 23–27, 2006), pp. 175–207.

[18] Kuhner. 20-L Apparatus. Basel, Switzerland: Kuhner AG; 2012. (Available at: http://safety. kuhner.com/en/product/apparatuses/safety-testing-devices/id-20-l-apparatus.html; last accessed November 3, 2012).

[19] Wu H-C, Kuo Y-C, Wang Y-H, Wu C-W, Hsiao H-C. Study on safe transporting velocity of nanograde aluminum, iron, and titanium. Journal of Loss Prevention in the Process Industries 2010;23:308–11.

[20] Boilard SP, Amyotte PR, Khan FI, Dastidar AG, Eckhoff RK. Explosibility of micron- and nano-size titanium powders. Krakow, Poland: Proceedings of Ninth International Symposium on Hazards, Prevention, and Mitigation of Industrial Explosions; July 22–27, 2012.

[21] CSB. Investigation report—sugar dust explosion and fire—Imperial Sugar Company. Report No. 2008-05-I-GA. Washington, DC: U.S. Chemical Safety and Hazard Investigation Board; 2009.

[22] Taveau J. Secondary dust explosions: how to prevent them or mitigate their effects? Process Safety Progress 2011;31:36–50.

Chapter 12

Myth No. 11 (Mixing): There's No Problem If Dust Is Not Visible in the Air

We are now at the halfway point in our examination of 20 dust explosion myths categorized by element according to the explosion pentagon. Much of the previous discussion has been focused, as might be expected, on the dust itself—i.e., the *fuel*. We have also considered *ignition source* and *oxidant* requirements, thus completing what is commonly known as the fire triangle. In this chapter we move on to the *mixing* element of the pentagon, followed by later chapters on *confinement* and concluding with several chapters on the pentagon in its entirety.

Mixing of sufficient fuel with an oxidizing atmosphere is needed to create an explosible fuel/air mixture. This is how a dust cloud is formed. So if a dust cloud is not visible in the open air, it may be tempting to think there should be no concern. The problem with this logic becomes apparent when considering the process plant locations where dust clouds exist and the mechanisms by which dust layers can be lofted into suspension. (See also Chapter 4.)

12.1 PRIMARY AND SECONDARY DUST EXPLOSIONS

Explosible dust clouds are optically thick. Eckhoff [1] puts this in practical terms by quoting the observation that a glowing 25-W light bulb cannot be seen through 2 m of a dust cloud at concentrations exceeding 40 g/m^3. To place this dust concentration in perspective, recall Section 4.5 and the suggested exercise in which representative values were given for minimum explosible concentrations of this magnitude and greater.

Initiation of dust explosions therefore usually happens in dust clouds present in process vessels and units such as mills, grinders, and dryers [1,2]—i.e., inside equipment where the conditions of the explosion pentagon are satisfied. (See Figure 3.9 and Section 3.2.) The reason for this occurrence is further explained by Figure 12.1. Here, the range of explosible dust concentrations in air at

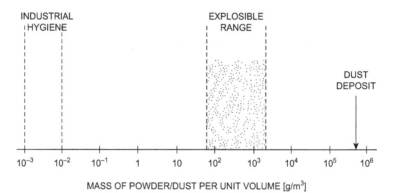

FIGURE 12.1 Typical dust concentration ranges of common natural organic dusts in air at normal temperature and pressure, for maximum permissible hygienic exposure, dust explosions, and dust deposit combustion, respectively [1].

normal temperature and pressure for a natural organic dust (e.g., cornstarch) is compared with the typical range of maximum permissible dust concentrations that are relevant in the context of industrial hygiene, and with a typical density of deposits or layers of natural organic dusts. Clearly, the range of explosible concentrations is orders of magnitude greater than the concentrations permitted in areas inhabited by workers.

Figure 12.1 also illustrates the importance of using an appropriate technique such as an explosion-proof vacuum for removing dust deposits from the work-place. Vigorous sweeping or cleaning with compressed air can readily create an explosible dust cloud by moving from right to left along the concentration axis in Figure 12.1. (See Section 4.3.)

Industry experience tells us that dust explosions also occur in process areas, not just inside process units. A *secondary* explosion can be initiated due to entrainment of dust layers by the blast waves arising from a *primary* explosion, as illustrated in Figure 12.2. The primary event might be a dust explosion originating in a process unit, or could be any disturbance energetic enough to

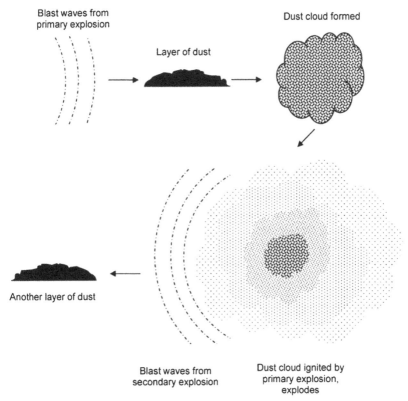

Blast waves from primary explosion

Dust cloud formed

Layer of dust

Another layer of dust

Blast waves from secondary explosion

Dust cloud ignited by primary explosion, explodes

FIGURE 12.2   Secondary dust explosion resulting from a primary explosion [3].

disperse combustible dust layered on the floor and other surfaces (such as the aforementioned cleaning activities).

An example of such an energetic disturbance (other than a primary dust explosion) is a gas explosion leading to a dust explosion. This is a well-documented phenomenon in the underground coal mining industry where devastating effects can result from the overpressures and rates of pressure rise generated in a coal dust explosion that has been triggered by a methane explosion (as in the case of the Westray mine explosion previously described in Chapter 4).

In their review of dust explosions in food industry plants, Spigno and De Faveri [4] cite an incident that occurred at a UK facility in which oil was being extracted from soy beans. Switching on a pump ignited a hexane leak, and the resulting gas explosion caused secondary dust explosions in the feed tunnels for the grain silos [4].

Taveau [5] has provided a comprehensive review of secondary dust explosion cases in France (Metz and Blaye) and the United States. The U.S. examples include the DeBruce grain elevator explosion as well as the major incidents investigated by the U.S. Chemical Safety Board (CSB). (See Chapter 3). He comments that secondary dust explosions are often more severe than the initiating event [5], and quotes (p. 36) Zalosh et al. [6], who give a succinct and clear account of the secondary dust explosion problem:

Perhaps the most devastating dust explosion scenario is the generation of a secondary dust explosion in the building surrounding the equipment in which a primary explosion takes place. The secondary explosion occurs when the blast wave emanating from the ruptured equipment or conveyor lifts the accumulated dust into suspension, and the flame from the primary explosion subsequently ignites the suspended dust cloud. The resulting devastation and casualties are associated both with the burning of building occupants and with the structural damage to the building.

Figure 12.3 illustrates an additional secondary explosion concern involving an outdoor dust filtration system that recycles process air back indoors. Davis et al. [7] highlight the need to consider the possibility of an explosion in the filter unit escalating back into the enclosed plant area and serving as the primary cause of a secondary explosion of dust accumulations.

There are therefore numerous primary/secondary sequences that require forethought through effective hazard identification and attention by means of appropriate risk reduction techniques. We have looked at the scenario of a gas explosion leading to a dust explosion, and that in which both explosions involve combustible dust. To conclude this section, we briefly examine the case of a gas explosion resulting in a dust flash fire.

The event in question is the third iron dust incident at the Gallatin, TN, Hoeganaes plant in which three workers died and two were injured [8]. (See Section 1.4.) The initial explosion was caused by hydrogen leaking into a process-pipe trench from an 8-cm (3-in.) by 18-cm (7-in.) hole in a vent pipe (see Figure 12.4). This primary explosion disturbed layered iron dust (such as shown

FIGURE 12.3   Filter installation located outdoors and recycling process air back indoors [7].

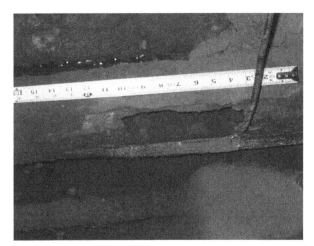

FIGURE 12.4   Hole in the hydrogen piping at the Hoeganaes Corporation facility in Gallatin, TN [8].

FIGURE 12.5 Iron dust combustibility demonstration performed by the U.S. Chemical Safety Board [8].

in Figure 1.4), which subsequently burned in suspension as a flash fire—the secondary event [9]. Figure 12.5 illustrates a laboratory-scale demonstration of a 30-g sample of iron dust dispersed above a 20-cm (8-in.) burner and producing a flame with a peak diameter of 46 cm (18 in.).

12.2 DOMINO EFFECTS

All of the primary/secondary events described in the preceding section are examples of what have come to be known in the process industries as *domino* or *knock-on* effects [10]. The avoidance of domino effects by application of inherently safer design principles [10–12] and advanced techniques such as Bayesian networks [13] is a topic of significant and current research interest throughout the world.

In their review of 261 incidents involving domino effects, Clini et al. [14] searched several databases including the Major Hazard Incident Data Service (MHIDAS). While dust explosions were not explicitly considered in their analysis, the most common primary incidents reported were fire (43%) and explosion (41%) [14]. The highest percentages of domino sequences determined were explosion/fire (21%), release/explosion/fire (15%), and fire/explosion (14%) [14]. Storage areas (37%) and process plants (27%) were the areas found to be most involved with domino incidents; 57% of the total incidents resulted in fatalities [14].

Again, the preceding statistics are for all hazardous materials, not just combustible dusts. Nevertheless, they give a good indication of the magnitude of the overall domino problem, of which dust explosions are an important subset. Reading the entry for the 1977 Westwego, LA, grain elevator disaster (see Table 3.1) in Spigno and De Faveri [4] (p. 126), one gets a sense of the devastation that can be brought about by domino effects:

Thirty-seven grain silos totally destroyed, others damaged by series of explosions. Cause not known but low humidity/pressure piling contributory factors. Fatalities caused when concrete tower fell on office block.

Further evidence of the extensive losses arising from these sequential events is given by Piccinini [15] in his description of a 2001 wool factory incident in Italy that caused considerable damage to the facility and killed three workers and injured five others. The domino chain in this case actually involved three distinct phases [15]:

- *Fire*—Electrical equipment overheated or sparked and caused an accumulation of vegetable particles to ignite, leading to smoldering or flaming combustion.
- *Explosion (primary)*—The fire burned nylon nets laden with vegetable dust and wool fibers, causing them to collapse and create a dust cloud that exploded.
- *Explosion (secondary)*—The primary explosion lofted and ignited dust accumulated on surfaces, nylon nets, and filter bags, resulting in multiple flamefronts and destructive overpressures.

Piccinini [15] calls the preceding sequence an *unusual* case of the domino effect. He reminds us, however, that the conditions leading to this chain of events are all too *usual*—all too familiar and common. It turns out that large quantities of combustible dust had been present in the complex workings and equipment for decades; all that was missing was an adequate ignition source [15].

12.3 REALITY

Potential dust explosion hazards exist in industry even without the observable presence of airborne dust. Operations involving plant personnel do not occur as a routine matter in environments with combustible dust clouds continuously present. Thus, *primary* dust explosions occur inside process units with *secondary* dust explosions arising in related work areas.

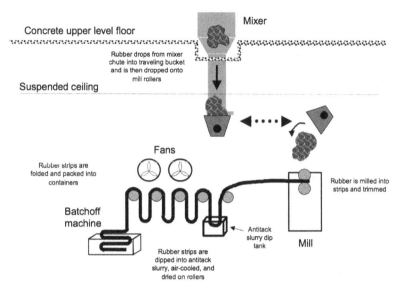

FIGURE 12.6 Simplified flow diagram of the two-story rubber compounding process at West Pharmaceutical Services in Kinston, NC [16].

Primary events other than an initial dust explosion can lead to secondary explosions or flash fires; these include gas explosions as well as general aerodynamic disturbances. Recalling Figure 4.3 and the accompanying discussion in Chapter 4, we know that relatively thin layers of dust deposited on surfaces can easily provide the fuel for a secondary dust explosion even if a primary dust explosion inside a process unit has been effectively excluded.

A relevant case study here is the dust explosion that occurred at West Pharmaceutical Services in Kinston, NC, on January 29, 2003, killing 6 people and injuring 38 others [16]. (See Section 3.1 and Section 4.3.) The plant manufactured rubber components for the health-care industry; in part of the operations, rubber was coated with polyethylene to prevent sticking. Figure 12.6 gives a simplified version of the rubber compounding process.

The coating process initially employed the concept of inherent safety (as subsequently described in Chapter 15) by running the rubber strips through a tank containing a slurry of polyethylene powder in water. After this stage, however, the water was evaporated by fans to leave only the required powder coating on the rubber strips. This drying step reintroduced the dust explosion hazard by the generation of airborne polyethylene dust that settled on surfaces above the suspended ceiling in the facility (Figure 12.6), and which therefore was not generally visible. Because of the extensive damage to the plant (see Figure 12.7), the U.S. Chemical Safety Board was unable to definitively determine the primary event that dispersed and ignited this dust to produce the devastating secondary explosion.

FIGURE 12.7   Explosion damage to the roof at West Pharmaceutical Services, with the two-story rubber compounding section of the facility in the background [16].

The following quotes from the CSB investigation report [16] (pp. 33 and 39, respectively) are particularly telling:

During interviews, West employees described no such visible dispersions of dust in the air. Thus, it is not a realistic probability that explosive concentrations were normally present in the production areas of the plant including the area around the batchoff machines.

and

CSB believes that the accumulation of combustible dust above the suspended ceiling is the most important safety issue in the West incident.

It should be noted that cleaning crews had continuously wiped and vacuumed dust from open surfaces so that the work area was essentially free of visible accumulation of dust [16].

12.4 WHAT DO *YOU* THINK?

Cozzani et al. [11] list three common features of incidents involving domino effects: (i) a primary scenario that initiates the domino sequence; (ii) an escalation vector that is generated by the physical effects of the primary event, propagates the primary event, and results in damage to at least one secondary equipment item; and (iii) one or more secondary scenarios involving the damaged equipment item(s). Table 12.1 summarizes these features based on an analysis of the aforementioned MHIDAS database [11].

The specific mechanisms for dust explosion and dust flash fire occurrence described in Section 12.1 can be shown to be consistent with the methodology behind the more general process safety considerations in Table 12.1. I encourage you to prove this by noting footnotes *b* and *c* to Table 12.1, as well as the explanation given by Cozzani et al. [11] that confined explosions include gas, vapor, and dust explosions inside vented or unvented equipment.

TABLE 12.1 Escalation Vectors and Expected Secondary Scenarios for
Various Primary Scenarios [11]

Primary scenario	Escalation vector	Expected secondary scenario[a]
Pool fire	Heat radiation, fire impingement	Jet fire, pool fire, BLEVE, toxic release
Jet fire	Heat radiation, fire impingement	Jet fire, pool fire, BLEVE, toxic release
Fireball	Heat radiation, fire impingement	Tank fire
Flash fire	Fire impingement	Tank fire
Mechanical explosion[b]	Fragments, overpressure	All[c]
Confined explosion[b]	Overpressure	All[c]
BLEVE (boiling liquid expanding vapor explosion)[b]	Fragments, overpressure	All[c]
VCE (vapor cloud explosion)	Overpressure, fire impingement	All[c]
Toxic release	–	–

[a]Expected scenarios also depend on the hazards of the target vessel inventory.
[b]Following primary vessel failure, further scenarios may occur (e.g., pool fire, fireball, toxic release).
[c]Any of the scenarios listed in the first column (primary scenario) may be triggered by the escalation vector.

To put Table 12.1 into practice, you could consider using it as the basis for a domino hazard analysis of your facility or part thereof. The first column can be used to generate what-if type questions, with the escalation vectors in the second column identifying the consequences of the primary event and the ultimate consequences as represented by the third column. As the case study presented by Piccinini [15] demonstrates, this type of analysis must be carried through until all consequences of an initiating event have been identified and remediated.

Have you considered domino incidents and their effects in your organization and at your facility?

REFERENCES

[1] Eckhoff RK. Dust explosions in the process industries, 3rd ed. Boston, MA: Gulf Professional Publishing/Elsevier; 2003.

[2] Amyotte PR, Eckhoff RK. Dust explosion causation, prevention and mitigation: an overview. Journal of Chemical Health & Safety 2010;17:15–28.

[3] Abbasi T, Abbasi SA. Dust explosions—cases, causes, consequences, and control. Journal of Hazardous Materials 2007;140:7–44.

[4] Spigno G, De Faveri DM. Safe design in dryers for food industry. Chemical and Biochemical Engineering Quarterly 2000;14:125–31.

[5] Taveau J. Secondary dust explosions: how to prevent them or mitigate their effects? Process Safety Progress 2011;31:36–50.

[6] Zalosh R, Grossel SS, Kahn R, Silva DE. Safely handle powdered solids. Chemical Engineering Progress 2005;101:23–30.

[7] Davis SG, Hinze PC, Hansen OR, van Wingerden K. Does your facility have a dust problem: methods for evaluating dust explosion hazards. Journal of Loss Prevention in the Process Industries 2011;24:837–46.

[8] CSB. Case study—Hoeganaes Corporation: Gallatin, TN—metal dust flash fires and hydrogen explosion. Report No. 2011-4-I-TN. Washington, DC: U.S. Chemical Safety and Hazard Investigation Board; 2011.

[9] Rigas F, Amyotte P. Hydrogen safety. Boca Raton, FL: CRC Press, Taylor & Francis Group; 2012.

[10] Kletz T, Amyotte P. Process plants. A handbook for inherently safer design, 2nd ed. Boca Raton, FL: CRC Press, Taylor & Francis Group; 2010.

[11] Cozzani V, Gubinelli G, Salzano E. Escalation thresholds in the assessment of domino accidental events. Journal of Hazardous Materials 2006;A129:1–21.

[12] Cozzani V, Tugnoli A, Salzano E. The development of an inherent safety approach to the prevention of domino accidents. Accident Analysis and Prevention 2009;41:1216–27.

[13] Khakzad N, Khan F, Amyotte P, Cozzani V. Domino effect analysis using Bayesian networks. Risk Analysis 2013;33:292–306.

[14] Clini F, Darbra RM, Casal J. Historical analysis of accidents involving domino effect. Chemical Engineering Transactions 2010;19:335–40.

[15] Piccinini N. Dust explosion in a wool factory: origin, dynamics and consequences. Fire Safety Journal 2008;43:189–204.

[16] CSB. Investigation report—dust explosion—West Pharmaceutical Services, Inc. Report No. 2003-07-I-NC. Washington, DC: U.S. Chemical Safety and Hazard Investigation Board; 2004.

Chapter 13

Myth No. 12 (Mixing): Once Airborne, a Dust Will Quickly Settle out of Suspension

Having examined several formation mechanisms for dust clouds in the preceding chapter, we now consider the fate of the particles once suspended in the oxidizing atmosphere. Although earlier discussion in the current book has cautioned against making broad, general comparisons between gas and dust explosions, there are, in fact, useful analogies that can be drawn between the two fuels to help in understanding some of the unique features of dusts.

Previous chapters have indicated that the mixing component of the explosion pentagon highlights a key, fundamental difference between gaseous and particulate fuel sources. Dusts are solids, and our experience tells us that solids are denser than gases. So we would reasonably expect the particles in a dust cloud to eventually settle out of suspension in the absence of a sustaining airflow. But should we expect all dusts to settle out "quickly," or even all at the same rate?

The answer is no—no more than we should expect all gases to behave in the same manner once dispersed into the atmosphere. Release parameters (e.g., pressure) and environmental conditions (e.g., wind speed and direction) will, of course, play a role in determining transport pathways and concentration levels, but the physical nature of the gas itself must also be a determinant factor.

How else then do we explain the layering of methane (molar mass of 16 g/mol) near the roof of the Westray mine (Chapter 4) [1]? Or the physics behind the personal account of J.K. Gehlawat [2] (p. 261), who arrived in Bhopal, India, by train at the time of the release of methyl isocyanate (molar mass of 57 g/mol) from the Union Carbide plant at midnight on December 2, 1984?

It was a horrifying experience to walk down the distance up to the Hotel since the heavy cloud of the pungent gas not only affected the respiratory system adversely due to inhalation but my eyes also became sore and red like it happens with a tear gas.... I knew that many gaseous reactants are highly soluble in water. Most are heavier than air and that on a chilly midnight the concentration of the toxic MIC [methyl isocyanate] would be much higher at the ground level than on the upper floors.

Among other factors to be discussed in the following sections, particle size and shape can be expected to figure prominently in determining the dispersibility (ease of dispersion) of a given dust. There is no *a priori* reason to expect similar dispersion behavior for the nm-size, near-spherical titanium sample, and the μm/mm-size, cylindrical-shaped polyamide 6.6 (nylon) sample, shown in the scanning electron microscope images in Figure 13.1.

13.1 DUSTINESS

In a recent conference presentation, Klippel et al. [5] described "a new safety characteristic—the dustiness, meaning the tendency of a dust to form clouds." This concept is based on VDI 2263-Part 9 [6] from the Association of German Engineers (VDI), which gives six dustiness groups; group

FIGURE 13.1   SEM micrographs of nano-size titanium (upper) [3] and flocculent polyamide 6.6 (lower) [4].

1 denotes dusts having minimal tendency to stay airborne; and group 6, the opposite [5].

Klippel et al. [5] identified the following parameters as influencing factors for dustiness (or dispersibility): (i) particle size, (ii) particle specific surface area, (iii) dust moisture content (humidity), (iv) dust density, (v) particle shape, and (vi) agglomeration processes resulting in larger-than-expected effective particle sizes. They used a vertical-tube apparatus to make concentration measurements for clouds formed by upward air dispersion of a dust layer resting on a porous filter plate [5]. Dustiness categorization results along with data for the first three of the aforementioned influencing factors and the explosibility parameters P_{max} and K_{St} are given in Table 13.1.

TABLE 13.1 Dustiness and Explosibility Parameters for Various Dusts [5]

Dust	Dustiness group	Median diameter [μm]	Specific surface area [m²/g]	Moisture content [%]	P_{max} [bar(g)]	K_{St} [bar·m/s]
Wheat flour	1	65	0.428	8.6	7.8	95
Sanding dust	2	260	0.874	4.3	7.6	84
Skimmed milk powder	3	45	0.207	3.7	7.6	117
Maize starch	4	14	2.486	4.5	8.7	167
Lignite	6	38	4.911	8.9	8.4	196
Potato starch	6	46	0.265	9.5	7.0	86

We will return to an examination of the general trends in Table 13.1 later in Section 13.6. For now, let us reason through some fundamentals of why dustiness is affected by the factors identified by Klippel et al. [5]:

- *Particle size*—For a given dust density, the terminal settling velocity of spherical particles increases with an increase in particle diameter [7].
- *Particle specific surface area*—Higher specific surface area leads to a lower settling rate because of greater drag force acting on the particles [5].
- *Dust moisture content (humidity)*—Cohesion caused by interparticle adhesion forces results in a decrease in dispersibility with an increase in dust moisture content [8].
- *Dust density*—For a given particle diameter, the terminal settling velocity of spherical particles increases with dust density [7].
- *Particle shape*—Features such as asymmetry in particle shape and roughness in surface texture have been shown to result in lower terminal settling velocities than for smooth, spherical particles due to rotational settling and eddy formation [9]. The suite of dusts tested by Klippel et al. [5] (Table 13.1) consisted of both spherical samples (e.g., potato starch) and flake-shaped samples (e.g., sanding dust). Flocculent or fibrous dusts would be expected to settle at rates dependent on the orientation of the cylindrical-shaped particles to both the flow and gravitational fields.
- *Agglomeration processes*—Two key aspects of the tendency of dusts to agglomerate become important when considering the concept of an effective particle diameter: (i) attraction between particles in dust layers due to interparticle cohesion forces, and (ii) rapid coagulation of particles in a dust cloud [8,10,11]. In the first instance, dispersion of agglomerates into primary particles is made more difficult; the second case means that even if a dust is well dispersed, the formation of larger agglomerates in suspension remains a possibility. Agglomerates (pre-dispersion) are clearly visible in some of the scanning electron microscope (SEM) images shown in the paper by Klippel et al. [5] —e.g., for skimmed milk powder and maize starch.

The preceding points draw in some measure on basic knowledge gained from dispersion studies of spherical particles undergoing gravitational settling. The flow situation in a well-mixed dust cloud of irregular-shaped, agglomerated particles, formed by an air blast of specific intensity, direction, and duration, will naturally involve more complex phenomena. These considerations, however, only serve to reinforce the fact that dusts do not necessarily settle out of suspension rapidly, nor do all dusts undergo the same settling process.

13.2 PREFERENTIAL LIFTING

Dust dispersion in the apparatus employed by Klippel et al. [5] was accomplished by a 2-second air pulse of flow rate 23 m^3/h emanating from beneath the dust (which, as previously mentioned, rested on a porous filter plate). Other orientations between the aerodynamic disturbance and the dust layer are, of course, possible. Of significant interest is the situation common in secondary dust explosions where the dust is lifted from a layer behind a propagating shock wave [12]. This is an important issue that has been the subject of both experimental and modeling research (e.g., Klemens et al. [13] and Zydak and Klemens [12], respectively).

A practical concern here is the settling of fine combustible dust on top of a layer of coarser material, or perhaps on top of a composite layer that has been rendered non-combustible by application of an inert dust. In either case, a layer of less-reactive or non-reactive dust would be essentially replaced by a thin layer of dust presenting a new explosion hazard. Preferential lifting of this top layer followed by ignition of the resulting cloud could have disastrous consequences.

This is a scenario that can unfold in many work locations, including underground coal mines containing float coal dust (–200 mesh or <75 μm) transported by ventilation air currents [14]. (The fact that float coal dust is so-named because it *floats* along in the ventilation air reminds us that both dust properties [e.g., particle size] and external mechanisms [e.g., forced convection] play a role in determining dust settling rates.) This fine dust might accumulate in mine areas remote from its source and thus pose a threat by settling on hard-to-clean overhead surfaces and also in terms of preferential lifting of dust layers.

The latter hazard involves float coal dust settling on a layer of mine dust freshly inerted with rock dust (as described in Section 7.1), thus effectively negating this prevention measure until the next application of rock dust. The potential magnitude of this problem in relation to rock dust inerting requirements is illustrated by Figure 13.2. Here, we see the total incombustible content or TIC (moisture and ash contents of the coal, as well as added rock dust) required to prevent explosion propagation in full-scale mine tests with various sizes of Pittsburgh seam high-volatile bituminous coal dust. The testing was conducted by the Pittsburgh Research Laboratory (PRL) of the U.S. National

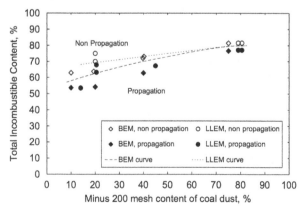

FIGURE 13.2   Effect of coal dust particle size on rock dust inerting requirements expressed as total incombustible content [15]. BEM, Bruceton Experimental Mine; LLEM, Lake Lynn Experimental Mine.

Institute for Occupational Safety and Health (NIOSH); up to 1996 the PRL had been part of the U.S. Bureau of Mines (USBM). The mines used were the older Bruceton Experimental Mine (BEM) and the newer, larger in cross-sectional area Lake Lynn Experimental Mine (LLEM) [15]. (See Figure 3.2 for a photograph of the Bruceton Experimental Mine.)

Reading from the LLEM data curve, approximately 70% TIC was needed for inerting of the 20% –200 mesh coal dust. The finer, 80% –200 mesh coal dust required about 80% TIC. This seemingly small increase in TIC from 70% to 80% actually corresponds to a large increase in inerting ratio (ratio of inert matter to combustible matter) from 2.3 to 4.0. (See Section 4.4.)

These data provide a basis for understanding the differentiated regulatory requirements for U.S. bituminous coal mines in effect at the time the paper by Sapko et al. [15] was published (2007). These requirements, based in part on mine dust sizes found in the 1920s, were at least 65% TIC in intake (nonreturn) airways and at least 80% TIC in return airways where there is a higher likelihood of float coal dust accumulation [15]. The data shown in Figure 13.2, coupled with the knowledge that modern mining methods have led to finer coal dust in intake airways than in the previous century, were employed by Sapko et al. [15] to make an eloquent argument for a higher TIC requirement in intake airways. It is of interest to note that current regulations for U.S. bituminous coal mines require at least 80% TIC in all underground areas [16].

13.3 NANO-MATERIALS

The concluding paragraphs of Section 8.3 and Section 11.4 dealt with recent findings concerning the ignitability and explosibility of dusts having particle sizes in the nano-range. While nano-dust explosion research is attracting increasing interest from industry, government, and academia [3,10,11,17–21], there is at present a limited amount of experimental data in this field.

As also mentioned previously in Section 8.3, the results to date seem to indicate that explosion severity is not significantly different at the nano-scale than at the micron-scale; however, the likelihood of an explosion increases significantly as the particle size decreases into the nano-range. Nano-scale materials are very sensitive and can self-ignite under the appropriate conditions during laboratory testing and handling (Dastidar, A.G., 2012. Personal communication, with permission). (See Section 11.4.)

Eckhoff [10,11] provides reasoning based on the agglomeration processes described earlier in Section 13.1 for the similar (or at least not greatly dissimilar) explosion severities (P_{max} and K_{St}) of micron- and nano-size samples of the same dust. Agglomerates can clearly be seen in the nano-titanium SEM image shown in Figure 13.1. Analysis of similar SEM pictures by Boilard et al. [3] showed some nano-agglomerates to be composed of about 50 primary particles while others were larger and made up of possibly thousands of particles. It would seem likely that such a powder would undergo dispersion leading to a cloud as shown on the right in Figure 13.3 as opposed to the one on the left.

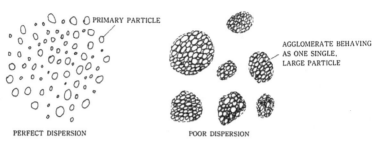

FIGURE 13.3 Schematic showing dust clouds formed by dispersion of primary particles (left) and agglomerates (right) [8,9].

The effective particle diameters in this case would be larger than those of the individual primary particles.

Eckhoff [10] speculates that the low minimum ignition energies of some nano-dusts may be due to the high temperatures found in electric spark plasmas. While it may be possible for particles in agglomerate form to maintain this shape when exposed to lower flame propagation temperatures, thermal stresses in the spark plasma region may facilitate the formation of a cloud of primary nano-particles from the larger agglomerates [10]. Eckhoff [10] further comments that increased turbulence leads to an enhancement of coagulation rates and therefore the formation of agglomerates from primary particles. Given the lower turbulence intensities found in typical MIE apparatus as compared to closed vessel explosion chambers used to determine P_{max} and K_{St}, one wonders if slower agglomeration processes during dispersion also lead to low MIEs for nano-size dusts.

The field of nano-dust explosibility is, at the time of writing this book, wide open and in need of further investigation from fundamental, experimental, and modeling perspectives. It would appear that at present we cannot say we fully understand the impact on dust explosibility of particle size reduction into the nanometer range because we do not know whether explosions of these materials are actually occurring in clouds of primary nm-size particles. From a pragmatic perspective, it would also be helpful to know whether such primary-particle clouds can be generated by actual industrial processes (Eckhoff, R.K., 2012. Personal communication, with permission).

13.4 FLOCCULENT MATERIALS

Chapter 2 makes several references to flocculent or fibrous materials. (See Figure 2.2, Figure 2.5, and the accompanying discussion.) These are dusts (also known as "flyings") that are better characterized by a length-to-diameter (L/D) ratio or an equivalent particle diameter, rather than an actual particle diameter as would be appropriate for spherical or near-spherical particles. Although flocculent materials have received some attention in the past (e.g., Bartknecht [22]), their explosion characteristics have also been the subject of focused research in recent years [4,21,23–27].

With respect to dispersibility, the nylon (polyamide 6.6) flock shown in the lower SEM image in Figure 13.1 might intuitively be expected to present a unique challenge when conducting closed-vessel explosibility testing. This is indeed the case with laboratory-scale chambers such as the 20-L vessel shown in Figure 11.7. Dispersion of large amounts (high concentrations) of flocculent dust from the external 1-L storage reservoir into the actual 20-L explosion chamber is made difficult for high L/D flock because of the low bulk density at this size. In their experimental work with nylon flock having a diameter of 19 μm and lengths up to 1 mm, Iarossi et al. [4] adopted the procedure of placing only up to 15 g of dust in the external 1-L reservoir. This corresponds to an eventual dust concentration of 750 g/m^3; for higher concentrations, the remainder of the sample mass was placed directly in the 20-L chamber.

The preceding paragraph describes the physical process of dispersing flocculent dust through an orifice/nozzle arrangement to form a dust cloud in an enclosed volume. But what of the coagulation processes that can lead to particle agglomeration upon dispersion? Marmo and Cavallero [24] describe a feature of flock production known as *activation* in which the fibers are colored and treated in a wet process that removes the sticky surfactant layer covering raw flock. Activated flock can be oriented by means of an electrostatic field and is easily suspended in air because the treated fibers do not stick together [24]. Marmo and Cavallero [24] determined that flock having the same dimensions, but different colors (meaning a different activation process), had different minimum ignition energies. One may speculate that this is due to both a chemical effect of the activation medium and the ease with which differently activated fibers are dispersed.

Armed with the fundamental material in the current chapter, we will now re-examine the wool factory incident previously presented in Section 12.2. This is the dust fire/primary dust explosion/secondary dust explosion sequence described by Piccinini [23] as a unique domino effect case. Our discussion on dust dispersion and dust/oxidant mixing places us in a better position to understand the description of this workplace as *the kingdom of dust* in which dust seemed to be permanently in suspension [23].

Further analysis of this domino chain and its consequences has recently been provided by Salatino et al. [26]. Figure 13.4 from their work shows SEM images of burr (lumps and flakes of wool mixed with vegetable bits found on sheep fleece) and noils (short wool fibers from textile processing) [26]. As described by Salatino et al. [26], a mixture of burr and noils (wool processing byproduct) was being charged to separation cells by pneumatic conveyance; gravity then separated the lighter, finer burr from the heavier, coarser noils, with the burr being dispersed and collecting on so-called venting nets and the noils settling out in deposits at the bottom of the cells. This description is consistent with that given by Piccinini [23] and helps to relate the concepts of dust dispersion and settling to actual industrial practice.

Also of note in this incident is the fact that wool is considered to be flame-resistant because of its relatively high ignition temperature. In practice then,

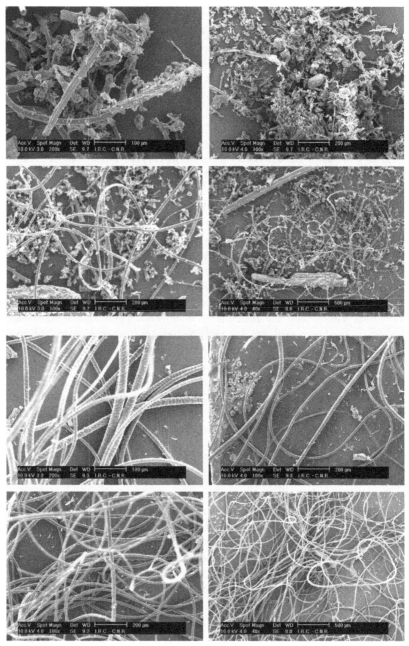

FIGURE 13.4 SEM micrographs of burr (upper grouping of four images) and noils (lower grouping of four images) [26].

wool dust fires and explosions may not be considered to be probable events during hazard identification and risk assessment exercises [26]. This calls to mind the discussion in Chapters 2 and 3 on whether and where dust explosions can occur, as well as the steps required before a given material can be judged with confidence to be non-explosible.

While even modest overpressures can be damaging to both people and structures, flame impingement and the associated high heat fluxes are leading causes of fatalities in industrial dust explosions (Zalosh, R., 2012. Personal communication, with permission). We therefore conclude this section with a brief look at the flammability of flocculent material.

The presence of strong air currents is believed to have aided the transition from initial dust layer smoldering combustion to flaming combustion in both the previously referenced wool factory [23,26] and another flock manufacturing plant [25]. This second incident, which also occurred in Italy in 2001, involved nylon fibers. Shutdown of a dryer allowed flock to settle on heated surfaces, resulting in nylon decomposition and the production of combustible off-gases [25]. One hypothesis of the explosion origin has flames being generated from the smoldering dust layer when the dryer fans were restarted; the flames are thought to have ignited the flammable atmosphere with the end result being dispersion of settled flock and a secondary dust explosion [25].

The flammability of nylon flock is clearly demonstrated in Figure 13.5 from the recent work of Iarossi et al. [4]. In the first and second photographs, we can observe spark ignition of the dust followed by flame propagation through the upper part of the glass tube. During these two steps, flame is generally visible only from about the electrode position upward through the tube. Some of the dust lies at the tube bottom, however, and is ignited by flashback of the primary flame. Hence, there is a secondary ignition and further propagation of flames, as shown in the third and fourth photographs in Figure 13.5. Afterward, a sticky, interwoven residue structure was found coating the interior of the glass tube.

Figure 13.5 demonstrates that flocculent material can be readily flammable and thus presents the additional hazard of being an ignition source for other materials present in flock-handling facilities. The result may be a further dust fire or explosion resulting in harm to personnel and damage to infrastructure.

13.5 REALITY

Numerous dust properties and external factors can affect the formation and duration of combustible dust clouds. Chief among these is dust particle size—not only of the primary, individual particles, but also the effective size of agglomerates either present pre-dispersion or formed post-dispersion. Other influencing factors include particle shape and specific surface area, and dust density and moisture content.

FIGURE 13.5   Explosion sequence for polyamide 6.6 (1.7 dtex and 0.5 mm length) in a MIKE 3 apparatus (1200 mg dust amount and 120 ms ignition delay time). From top to bottom: primary ignition, primary flame propagation, secondary ignition, and secondary flame propagation [4].

Fine, micron-size particles can experience preferential lifting over underlying dust when subjected to an aerodynamic disturbance. Additionally, recent research has revealed some of the unique issues presented by nano-size and flocculent dusts in terms of their dispersibility and explosibility. The field of nano-dust explosions, in particular, is in need of concerted research efforts.

The overall conclusion to be drawn from this chapter is that all dusts will settle given time and opportunity—they just won't all do it at the same rate.

13.6 WHAT DO *YOU* THINK?

As promised in Section 13.1, we now return to Table 13.1 for a closer look at the influence of the various dust properties on the reported dustiness. Recall that moving from group 1 to group 6 means an increasing likelihood of the particles remaining in suspension [5].

FIGURE 13.5—Cont'd

As a guide for the following exercise, you might consider a *lower* value of dustiness to be 1–3 and a *higher* value as 4–6. For the properties of the dust (*smaller* and *larger* median diameter, *smaller* and *larger* specific surface area, and *lower* and *higher* moisture content), a relative comparison within the given property data set is appropriate. Note that we are searching for specific cases where the dust property influence on dustiness is in accordance with the expectations described in Section 13.1.

Looking first at particle size as expressed by the median diameter, find an example in Table 13.1 of a larger diameter and a lower dustiness group. Now look for a smaller diameter and a higher dustiness group. Do the same for specific surface area (i.e., smaller area and lower dustiness; larger area and higher dustiness) and for moisture content (i.e., higher moisture and lower dustiness; lower moisture and higher dustiness).

Let us now examine the overall trends displayed by the data in Table 13.1. What are your observations in this regard? Would you agree that while the change in dustiness with variation in a given property is generally in accordance with expectations, exceptions do exist? That the expected trend arguably appears more prominently for specific surface area than for median diameter and moisture content? Or do you see something different in the data?

As a broad conclusion, there must be an interplay among the influencing parameters (both those shown in Table 13.1 and those not shown) that bears further investigation. For example, the potato starch sample has the highest moisture content, second-smallest specific surface area, and third-largest median diameter; yet it is in the highest dustiness group. Of interest is the potato starch SEM picture given by Klippel et al. [5], which displays a mix of near-spherical primary particles and some agglomerates. Could it be that the primary particles remain as separate entities and the agglomerates are effectively broken up during dispersion?

Further work in this important area is clearly required. This point is well recognized by Klippel et al. [5], who report their intention to pursue an investigation of industrially relevant dispersion methods at both laboratory and industrial scales. A parametric study in which attempts are made to isolate the influence of each dust property/dispersion factor would seem highly beneficial. Would it also be helpful to include an indication of the actual particle size range—say, a measure such as polydispersity [28]? (See Section 6.4.)

I would not draw any firm conclusions on the relationship between dustiness and explosion severity based solely on the P_{max}/K_{St} data given in Table 13.1. While there is clearly a relationship between the dispersibility of a dust and its explosibility, these P_{max}/K_{St} values would have been determined by means of standardized, closed-vessel testing in which a high-pressure air blast is employed for dust dispersion. (See Section 8.2. and Chapter 19.)

To conclude, reference is made to a practical application of the dustiness concept. Note the following statement appearing as part of a question recently posed to an online expert opinion service [29]: "Graphite powder seems to prefer an airborne state rather than settling down." What are your thoughts on this comment in light of the discussion in the current chapter? What factors would tend to keep a powder in suspension?

The follow-on question was [29]: "How thick does the cloud need to be before the place blows up?" The response is interesting because it nicely addresses the fact that explosible dust clouds are optically thick, as described previously in Section 12.1 [29]: "Generally speaking the MEC [minimum explosible concentration] of dust clouds is rarely below about 30 g/m^3. Dust clouds with a concentration above the MEC would look like a thick fog and one wouldn't be able to see objects a few feet away."

REFERENCES

[1] Richard KP, Justice. The Westray story—a predictable path to disaster. Report of the Westray Mine Public Inquiry. Halifax, NS, Canada: Province of Nova Scotia; 1997.

[2] Gehlawat JK. Bhopal disaster—a personal experience. Journal of Loss Prevention in the Process Industries 2005;18:261–3.

[3] Boilard SP, Amyotte PR, Khan FI, Dastidar AG, Eckhoff RK. Explosibility of micron- and nano-size titanium powders. Krakow, Poland: Proceedings of Ninth International Symposium on Hazards, Prevention, and Mitigation of Industrial Explosions; July 22–27, 2012.

[4] Iarossi I, Amyotte PR, Khan FI, Marmo L, Dastidar AG, Eckhoff RK. Explosibility of polyamide and polyester fibers. Krakow, Poland: Proceedings of Ninth International Symposium on Hazards, Prevention, and Mitigation of Industrial Explosions; July 22–27, 2012.

[5] Klippel A, Scheid M, Krause U. Investigations into the influence of dustiness on dust explosions. Krakow, Poland: Proceedings of Ninth International Symposium on Hazards, Prevention, and Mitigation of Industrial Explosions; July 22–27, 2012.

[6] VDI. VDI 2263-Part 9. Determination of dustiness of bulk material. Verein Deutscher Ingenieure. Available at http://www.vdi.de/4080.0.html?&tx_vdirili_pi2[showUID]=92714; 2008. (last accessed October 30, 2012).

[7] Green DW, Perry RH, editors. Perry's chemical engineers' handbook. 8th ed. New York, NY: McGraw-Hill; 2007.

[8] Eckhoff RK. Understanding dust explosions. The role of powder science and technology. Journal of Loss Prevention in the Process Industries 2009;22:105–16.

[9] Eckhoff RK. Dust explosions in the process industries, 3rd ed. Boston, MA: Gulf Professional Publishing/Elsevier; 2003.

[10] Eckhoff RK. Does the dust explosion risk increase when moving from μm-particle powders to powders of nm-particles? Journal of Loss Prevention in the Process Industries 2012;25:448–59.

[11] Eckhoff RK. Influence of dispersibility and coagulation on the dust explosion risk presented by powders consisting of nm-particles. Krakow, Poland: Proceedings of Ninth International Symposium on Hazards, Prevention, and Mitigation of Industrial Explosions; July 22–27, 2012.

[12] Zydak P, Klemens R. Modelling of dust lifting process behind propagating shock wave. Journal of Loss Prevention in the Process Industries 2007;20:417–26.

[13] Klemens R, Zydak P, Kaluzny M, Litwin D, Wolanski P. Dynamics of dust dispersion from the layer behind the propagating shock wave. Journal of Loss Prevention in the Process Industries 2006;19:200–9.

[14] Cashdollar KL. Coal dust explosibility. Journal of Loss Prevention in the Process Industries 1996;9:65–76.

[15] Sapko MJ, Cashdollar KL, Green GM. Coal dust particle size survey of US mines. Journal of Loss Prevention in the Process Industries 2007;29:616–20.

[16] Harris ML, Sapko MJ, Varley FD, Weiss ES. Coal dust explosibility meter evaluation and recommendations for application. Information Circular (IC) 9529. Pittsburgh, PA: National Institute for Occupational Safety and Health; 2012.

[17] Holbrow P. Explosion properties of nanopowders. Manchester, UK: Hazards XXI, IChemE Symposium Series No. 155; November 10–12, 2009. pp. 70–78.

[18] Bouillard J, Vignes A, Dufaud O, Perrin L, Thomas D. Ignition and explosion risks of nanopowders. Journal of Hazardous Materials 2010;181:873–80.

An Introduction to Dust Explosions

[19] Amyotte PR. Are classical process safety concepts relevant to nanotechnology applications? Journal of Physics: Conference Series (Nanosafe2010: International Conference on Safe Production and Use of Nanomaterials) 2011;304: 012071.

[20] Vignes A, Munoz F, Bouillard J, Dufaud O, Perrin L, Laurent A, Thomas D. Risk assessment of the ignitability and explosivity of aluminum nanopowders. Process Safety and Environmental Protection 2012;90:304–10.

[21] Worsfold SM, Amyotte PR, Khan FI, Dastidar AG, Eckhoff RK. Review of the explosibility of nontraditional dusts. Industrial & Engineering Chemistry Research 2012;51:7651–5.

[22] Bartknecht W. Dust explosions. Course, prevention, protection. Berlin: Springer-Verlag; 1989.

[23] Piccinini N. Dust explosion in a wool factory: origin, dynamics and consequences. Fire Safety Journal 2008;43:189–204.

[24] Marmo L, Cavallero D. Minimum ignition energy of nylon fibres. Journal of Loss Prevention in the Process Industries 2008;21:512–17.

[25] Marmo L. Case study of a nylon fibre explosion: an example of explosion risk in a textile plant. Journal of Loss Prevention in the Process Industries 2010;23:106–11.

[26] Salatino P, Di Benedetto A, Chirone R, Salzano E, Sanchirico R. Analysis of an explosion in a wool-processing plant. Industrial & Engineering Chemistry Research 2012;51:7713–18.

[27] Amyotte PR, Cloney CT, Khan FI, Ripley RC. Dust explosion risk moderation for flocculent dusts. Journal of Loss Prevention in the Process Industries 2012;25:862–9.

[28] Castellanos D, Carreto V, Mashuga C, Trottier R, Mannan SM. The effect of particle size dispersity on the explosibility characteristics of aluminum dust. Krakow, Poland: Proceedings of Ninth International Symposium on Hazards, Prevention, and Mitigation of Industrial Explosions; July 22–27, 2012.

[29] Ebadat V. Graphite powder seems to prefer an airborne state rather than settling down. How thick does the cloud need to be before the place blows up? Combustible Dust & Static Electricity Q&A. Chilworth. Available at http://pbs.canon-experts.com/expert/dr-vahid-ebadat/; 2012 (last accessed October 29, 2012).

Chapter 14

Myth No. 13 (Mixing): Mixing Is Mixing; There Are No Degrees

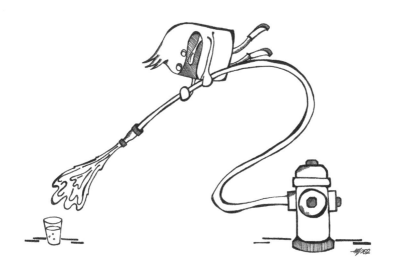

The subject of this chapter—turbulence—permeates the entire book and is therefore implicit in all previous chapters and those that follow. Tamanini [1] (p. 9) remarks that, with respect to modeling of dust explosions, "turbulence is the single most important factor whose effects need to be incorporated." Amyotte et al. [2] (p. 293) open their review paper on effects of turbulence on dust explosions with the comment that "an inherent part of the dust explosion problem is understanding and dealing with the influence of turbulence on explosion development." So we have a general picture of turbulence being both important and inextricably linked to the topic of dust explosions.

When discussing the different mixing requirements for gaseous and solid fuels in Chapter 5, I wrote that since dust particles are strongly influenced by gravity, an essential prerequisite for a dust explosion is the formation of a dust/oxidant suspension via adequate mixing. I then commented that this means there must be some level of turbulence in a dust cloud. Barton [3] makes the same point by indicating the presence of some degree of turbulence in a dust cloud because of the air movement needed for cloud formation.

Notice in the preceding paragraph that the word used is *some*, not *same*, to describe the level or degree of turbulence existing in a dust cloud. Turbulence is indeed ubiquitous in the world of dust explosions; it is always there either in the background as a factor to be acknowledged, or front and center as a parameter being manipulated by process conditions. But this does not mean that turbulence is a universal constant in the world of dust explosions. It is, in fact, a variable that must be recognized as having a significant impact on the mixing that occurs in a dust cloud both pre- and post-ignition.

14.1 TURBULENCE

Attempting to describe turbulence in a few paragraphs with no equations is a daunting task. I have therefore relied on classical, authoritative treatments of turbulence from chemical engineering texts [4–6] and a fundamental review paper [7]. The descriptions that follow (with underlining added for emphasis) are intentionally brief and are intended to demonstrate the connection between turbulence and mixing. (See also Section 14.4 for some additional thoughts.)

Holland and Bragg [4] (p. 5) describe turbulent flow as a "chaotic, fluctuating type of motion that promotes <u>rapid mixing</u>" on a scale commensurate with the geometry and dimensions of the flow environment. Coulson et al. [5] (pp. 700–701) further characterize turbulent flow as having a "complex interconnected series of circulating or eddy currents responsible for <u>fluid mixing</u>." Closer to the subject at hand, Eckhoff [7] (p. 108) writes: "In practical terms, turbulence is a state of <u>rapid internal, more or less random movement</u> of elements of the dust cloud relative to each other in three dimensions."

Eckhoff [7] also describes the impact of turbulent mixing on dust cloud ignitability and explosiblity. The likelihood of ignition by a hot surface, electric spark, or some other source is affected by the turbulent removal of heat from the

ignition zone; this means that ignition temperature and ignition energy requirements generally increase in magnitude at higher turbulence levels [7] (leading
to what could be termed an enhanced state of *mixedness*). Upon ignition, greater
mixedness results in what Eckhoff [7] refers to as a dynamic, three-dimensional
structure consisting of zones of burned, burning, and unburned particles. The
net effect is an increase in the number of combustion sites and hence overall
flame surface area, thus leading to more rapid combustion and overpressure
development at higher turbulence levels.

   The thoughts in the preceding paragraph can also be expressed using a general transport phenomena approach. From this perspective, greater fluid mixing caused by turbulence leads to higher rates of heat, mass, and momentum
transfer [5] due to steeper gradients in the respective driving force elements of
temperature, mole fraction, and velocity [6].

   You may have noticed my use of the words *level* and *degree* to denote a state
of turbulence. While this is common in the dust explosion literature, it is a short-
cut of sorts that blends the two main features of turbulence: scale and intensity.
These turbulence characteristics can be described, respectively, as being a measure of mean eddy size and as a function of fluid circulation velocity within the
eddies [6]. Both scale and intensity are variable due to, again respectively, (i) the
existence of different eddy sizes in a turbulent flow field, and (ii) superimposing
of a random velocity component (arising from the aforementioned fluid circulation within eddies) on the steady-state mean velocity [6].

   These considerations become increasingly complicated with turbulent dust
combustion given that the aerodynamic disturbances responsible for dust cloud
formation are often transient in nature. In this case, the occurrence of non-
stationary flow essentially negates the concept of a steady-state mean velocity,
and the issue becomes one of attempting to measure [8] and model [9] disper-
sion-induced turbulence decay. The matter increases in complexity yet again
when one realizes that such measurements and models relate to the turbulent
gas flow field in which *dust* particles of different sizes and shapes may not be
able to follow the velocity fluctuations.

   As a brief example, consider Figure 14.1, which shows the initial air flow
patterns in a standard 20-L explosion chamber for three different dispersion

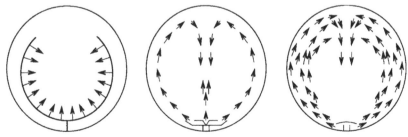

FIGURE 14.1   Initial direction of dispersion air flow in a standard 20-L chamber for the perforated ring nozzle (left), rebound nozzle (center), and Dahoe nozzle (right) [10].

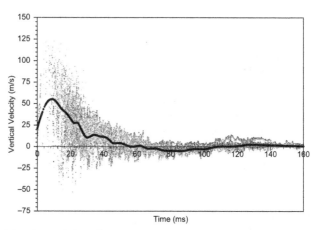

FIGURE 14.2   Time variation of instantaneous vertical velocity and extracted mean flow near the geometric center of a standard 20-L chamber for ten overlaid dispersion tests using the Dahoe nozzle [8].

nozzles. (Of the three, the rebound nozzle is most widely used in 20-L chambers throughout the world, having essentially replaced the perforated ring nozzle because of dust plugging issues with the annular holes.) The actual turbulent flow inside the vessel is, of course, more detailed than can be illustrated schematically. Figure 14.2, however, provides a glimpse of the chaotic, random nature of turbulence.

Here, we see clear evidence of the rapidly decaying flow field during the pre-ignition period in a standard 20-L chamber. The appearance of negative velocities in Figure 14.2 is due to the use by Mercer et al. [8] of a measurement technique known as Laser Doppler Anemometry (LDA), making it possible to resolve the downward (positive) and upward (negative) components of the velocity vector. The predominance of positive velocities in Figure 14.2 is consistent with the initial downward flow pattern in the chamber center for the Dahoe nozzle shown in Figure 14.1.

For our purposes here, we can leave these complications at the awareness stage and proceed on a more general basis to examine the influence of turbulence on the key parameters that define dust explosion likelihood and severity.

14.2 INFLUENCE OF TURBULENCE

Preliminary indication of the influence of turbulence on dust explosions was given in the preceding section in terms of increased ignition requirements and heightened combustion rates. Before examining further how these effects manifest themselves with respect to the corresponding explosion parameters given in Table 5.1, we need to distinguish between pre- and post-ignition turbulence [2,7,11].

As the name implies, pre-ignition, or initial, turbulence is the turbulence level existing in a dust cloud during its formation and prior to ignition. The coverage in Chapters 12 and 13 on primary dust/air suspensions inside process units and secondary dust clouds formed by lofting of layered dust is relevant here. Another example is the turbulent dust/air mixture created during the dispersion process in a closed or vented system during standardized explosibility testing; dispersing air pressure and ignition delay time are the two factors that primarily determine the pre-ignition turbulence level in this case.

Post-ignition, or explosion-generated, turbulence is, again as the name implies, the additional turbulence created in the unburned portion of a dust cloud after ignition and initial flame propagation. This topic is treated separately in Chapters 16 and 17.

Strictly speaking, for a given dust sample, there is only one value of each of MIE, P_{max}, and K_{St}—all as determined in standardized test apparatus under standardized conditions. This means that if standardized MIE determination is being conducted using a MIKE 3 apparatus, the dispersing air pressure will be set at 7 bar(g), and the ignition delay time will be varied incrementally with typical values of 60, 90, 120, 150, and 180 ms. As delay time is increased, turbulence level decreases, and the apparent MIE observed at a lower delay time may, in fact, decrease (assuming no concentration effects as discussed in the next section).

Similarly, if standardized P_{max}/K_{St} determination is being conducted using a Siwek 20-L chamber, the dispersing air pressure will be set at 20 bar(g) and the ignition delay time at 60 ms. Alteration of either of these test parameters will change the pre-ignition turbulence level (as per Figure 14.2) and will therefore change the measured values of explosion overpressure and rate of pressure rise. The effect on these explosion parameters of varying the ignition delay time is clearly shown in Figures 14.3 and 14.4, respectively. Increasing the ignition delay time results in lower values of both parameters.

As indicated earlier, a particular sample of dust would have only one value of P_{max} and K_{St} according to standardized terminology Although this was not the objective of the work of Sanchirico et al. [12], from a standardized testing approach, each data point in Figures 14.3 and 14.4 would correspond to the maximum overpressure and maximum rate of pressure rise (with unit conversion), respectively, from the pressure/time trace at the specified dust concentration and ignition delay time. (See Chapter 19.)

Pre-ignition turbulence is seen to have a much greater effect on rate of pressure rise (Figure 14.4) than on explosion overpressure (Figure 14.3). This is typically explained by the reasoning that overpressure is a thermodynamic parameter and hence is dependent largely on the initial and final conditions of the reacting mixture. Explosion overpressure is not entirely immune to the effects of turbulence, however, because highly turbulent closed-vessel dust explosions lead to more adiabatic-like conditions and hence lower heat losses. Rate of pressure rise, on the other hand, is a kinetic parameter and is therefore significantly affected by mixture conditions en route to completion of reaction.

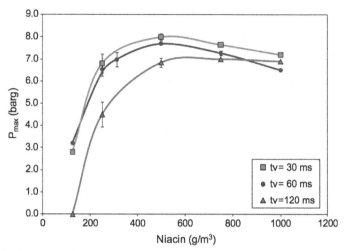

FIGURE 14.3 Influence of ignition delay time on the explosion overpressure of niacin dust in a Siwek 20-L chamber with a 10-kJ ignition energy [12].

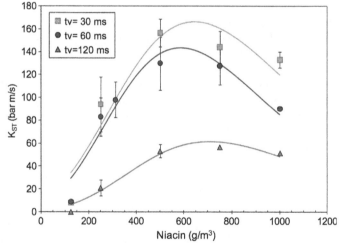

FIGURE 14.4 Influence of ignition delay time on the rate of pressure rise of niacin dust in a Siwek 20-L chamber with a 10-kJ ignition energy [12].

We conclude this section with a note on explosion relief venting and the importance of P_{max}/K_{St} measurement under standardized conditions of turbulence. As noted by Barton [3], standardized tests for these parameters are the basis for determining vent size requirements according to appropriate codes and standards. He further comments that the turbulence level in these laboratory-scale tests is largely representative of those encountered on a larger scale in most industrial applications [3]. Tamanini [1] refers to these conditions as "near worst-case," which is consistent with Barton's note that vent areas calculated using standardized test data are generally effective in all but the most extreme cases [3].

The take-away lesson from the preceding paragraph is that accepted vent sizing equations rely on input data for P_{max} and K_{St} determined according to standardized test protocols. This automatically implies a specified turbulence level that, if altered, invalidates both the experimental test procedure and the code- or standard-derived vent area.

Having said this, the relationship between turbulence and venting has been one of the most debated and intensively researched topics in the dust explosion literature over the years. The discussion often concerns whether current vent-sizing protocols are overly conservative for industrial scenarios involving lower turbulence levels than that employed in standardized testing, or whether these protocols underestimate vent areas for dusts with low K_{St} values or industrial scenarios with significant levels of explosion-generated turbulence. For readers interested in learning more on this subject, I recommend the excellent summaries by Tamanini [1] for developments up to 1998 and by Zalosh [13] covering the period up to 2008.

14.3 CONCENTRATION GRADIENTS

Section 14.1 highlighted the fact that turbulence is intimately tied to mixing—the mixing of solid fuel particles in a turbulent gas flow field. Particle size and shape and all of the other dustiness influencing factors described in Section 13.1 play a key role in the effectiveness of this mixing process. When one thinks of particle size not as a single value but as a distribution of sizes (diameters, effective diameters, L/D ratios, etc.), it becomes apparent that finer particles may become segregated from coarser particles. (Recall the wool factory incident described in Sections 12.2 and 13.4.) Additionally, we saw in Chapter 12 that dust dispersion events arise in many different forms.

It is therefore not surprising to read statements in the technical literature to the effect that local dust concentrations vary irregularly with time because of the irregular particle movement in a turbulent dust cloud [14]. Or that homogeneous dust cloud generation usually does not occur as a result of the typical cloud formation mechanisms of layer dispersion, dust conveyance, and enclosure filling [15]. Or that data trends such as those depicted in Figure 14.3 and especially Figure 14.4, although certainly caused by reduced turbulence levels at higher ignition delay times, must also be influenced by dust sedimentation at these higher delay times [12].

The last sentence in the preceding paragraph is yet another reminder of the importance of adopting standardized procedures for closed-vessel explosibility testing. The objective for dust dispersion in these tests is the formation of a highly turbulent, homogeneous (uniform concentration) dust cloud. In general, this can be accomplished [8,16]—but only under the appropriate conditions of dispersing air pressure and ignition delay time and with an appropriate dispersion system including the nozzle.

Further evidence of the link between turbulence and potential concentration gradients in dust clouds is obtained by comparing Figures 14.4 and 14.5. Figure 14.5, again from the interesting work of Sanchirico et al. [12], shows rate of pressure rise data for gaseous acetone in the form of K_G, the gas-phase equivalent of K_{St}, as explained in Chapter 19. (In accordance with the discussion in Section 14.2, Figure 14.5 also reports a single value of K_G for acetone from standardized testing reported by the U.S. National Fire Protection Association, or NFPA.)

With the exception of there being no dispersion in the initially quiescent tests for acetone, identical dispersion and ignition conditions were used to generate the data in Figures 14.4 and 14.5 [12]. Thus, the differences in rate of pressure rise for acetone at the various ignition delay times (0, 30, 60, and 120 ms) should be due solely to differences in turbulence level. Again, this conclusion would not hold for the niacin dust rate of pressure rise variation with delay time because of the possible influence of concentration differences.

The same point has been made with respect to explosions in interconnected vessels by Holbrow et al. [17], who comment that lean and rich zones can be expected to occur in turbulent dust/air mixtures because of the generation of concentration gradients. They also provide a mechanistic explanation for the dust particles not always being able to follow the flow patterns found in the turbulent eddies [17]. Kosinski and Hoffmann [18] make a similar argument concerning the non-uniform distribution of dust in interconnected vessels, particularly the secondary vessel into which flames are propagating. We

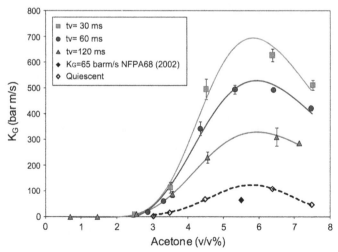

FIGURE 14.5 Influence of ignition delay time on the rate of pressure rise of acetone in a Siwek 20-L chamber with a 10-kJ ignition energy [12].

return briefly to this scenario of sequential dust explosions in interconnected vessels in Chapter 17.

14.4 REALITY

The level or degree of turbulence in a dust cloud varies according to the mode of cloud generation—a strong air blast, a less energetic aerodynamic disturbance, air movement devices inside process units, etc. Since turbulence is directly related to mixing effects, and turbulence can vary both spatially and temporally within a dust cloud, local variations in dust concentration may also occur.

Awareness of the effects of turbulence on dust explosions is critical from at least two perspectives. First, with respect to industrial practice, it is important to recognize that an increase in turbulence level leads to both heightened ignition requirements and explosion consequences. In other words, raising the pre-ignition turbulence level in a dust cloud will make it harder to ignite but also prone to more rapid combustion and thus higher overpressures and (especially) rates of pressure rise.

Second, with respect to the acquisition and use of explosion parameters such as MIE, P_{max}, and K_{St}, it is important to recognize that standardized test equipment employs a specific level of turbulence that optimizes dust concentration uniformity to the greatest extent possible. Departures from these standardized test conditions are ill advised, as would be the use of data so determined.

Given the prominence of turbulence considerations in my own doctoral research [2,19], I would like to conclude this summary section by describing what I consider to be a watershed event in dust explosion research. As a relatively new PhD student, I was fortunate enough to attend the 20th International Symposium on Combustion held at the University of Michigan in the United States. There, I witnessed the presentation by Professor Bill Kauffman of his paper titled "Turbulent and Accelerating Dust Flames [20]." (This is the same Bill Kauffman as mentioned in Section 3.1.)

Perhaps it was just my seeking to be inspired in my own research, but I do recall a definite "buzz" in the presentation room. This was the first time, to the audience's collective knowledge, that anyone had directly and quantitatively measured the pre-ignition turbulence intensity in a dust explosion chamber. Kauffman et al. [20] used a technique known as Hot Wire Anemometry (HWA) to measure turbulence intensities in the cold flow without dust particles (i.e., during air-only dispersion tests). They overcame the problem of decaying turbulence by means of fast-acting valves on the inlet and exhaust of their 1-m^3 spherical combustion vessel; these valves were closed at the moment of ignition to provide a constant volume having a near uniform turbulence level [2].

Dust explosion researchers could now observe the impact of measured turbulence intensities on overpressure and rate of pressure rise development for cornstarch/air suspensions. P_{max} increased by more than 50%, and $(dP/dt)_{max}$ more than doubled with an increase in turbulence intensity from 1.5 m/s to 4.2 m/s [2].

Subsequent years have seen several research groups make turbulence measurements in standard 20-L and 1-m³ chambers as well as vessels having other volumes and geometries. It appeared at one time that a fruitful line of research would be to make similar measurements in industrial-scale units and compare with the smaller-scale results. To my knowledge, this has not been undertaken to any great extent.

Rather, it seems that the laboratory-scale (especially 20-L) turbulence measurements have found valuable use in computational fluid dynamic (CFD) models to relate laminar burning velocity, S_L, to its turbulent counterpart, S_T [9]. (As noted by Skjold [9], laminar and turbulent burning velocities denote the flame propagation velocity relative to the unreacted mixture under laminar and turbulent flow conditions, respectively.) Zalosh [13] remarked that with continued development and validation, such CFD models will be helpful in understanding the effects of non-standard conditions of dispersion and turbulence in vented dust explosions.

14.5 WHAT DO *YOU* THINK?

Various dust handling units and dust explosion scenarios are therefore associated with different turbulence levels. What qualitative label for turbulence (say, *low* or *high*) would you place on each item in the following list?

- Gravity filling of a process vessel with coarse dust
- Baghouse operation with high dust/air suspension velocities
- A dryer with fans inducing strong forced convection currents
- Gentle sweeping of a dust layer
- A blast of high-pressure air from a hose aimed at a dust layer
- A secondary dust explosion in a work area caused by a severe primary explosion
- Standardized testing for minimum ignition energy
- Dust explosion propagation through ductwork connected to a process vessel
- Powder conveyance through a pipe by means of a screw-feed system
- Standardized testing for maximum explosion pressure and size-normalized maximum rate of pressure rise.

I get four *lows* and six *highs*. But then again, I made up the list, so I know exactly what I meant by each item. Perhaps we should add *moderate* to the label choices to give more of a sense of a wide range of possibilities.

Sometimes when attempting to explain a very broad concept in my university teaching, I have found it helpful to refer to a basic dictionary definition of the topic in question. Here is what the Merriam-Webster Online Dictionary has to say about *turbulence*—"the quality or state of being turbulent: as (a) great commotion or agitation, (b) irregular atmospheric motion especially when characterized by up-and-down currents, (c) departure in a fluid from a smooth flow."

All three meanings certainly fit with our discussion here on turbulence. The reference to up-and-down currents reminds me of the traditional nursery rhyme, "The Grand Old Duke of York:"

The grand old Duke of York,

He had ten thousand men.

He marched them up to the top of the hill,

And he marched them down again.

And when they were up, they were up,

And when they were down, they were down,

And when they were only halfway up,

They were neither up nor down.

I have fond memories of my young children singing and going through the actions of standing up and sitting down (fully and halfway), with multiple verses of increasing tempo. The end result was invariably Holland and Bragg's "chaotic, fluctuating type of motion [4]" often followed by a complete system collapse. This may be a bit much to try in your next training session on the effects of turbulence on dust explosions, but who knows?

Do you have other non-technical examples that help explain technical topics such as turbulence? Do we need more of these common, everyday examples to help us communicate our thoughts on technical issues to all stakeholders, not just those with the requisite technical background? (You can likely tell from the manner in which I phrased this question that I believe the answer is a resounding *yes*.)

REFERENCES

[1] Tamanini F. The role of turbulence in dust explosions. Journal of Loss Prevention in the Process Industries 1998;11:1–10.

[2] Amyotte PR, Chippett S, Pegg MJ. Effects of turbulence on dust explosions. Progress in Energy and Combustion Science 1988;14:293–310.

[3] Barton J, editor. Dust explosion prevention and protection. A practical guide. Rugby, UK: Institution of Chemical Engineers; 2002.

[4] Holland FA, Bragg R. Fluid flow for chemical engineers, 2nd ed. London, UK: Edward Arnold; 1995.

[5] Coulson JM, Richardson JF, Backhurst JR, Harker JH. Coulson and Richardson's chemical engineering volume I—fluid flow, heat transfer and mass transfer, 6th ed. Oxford, UK: Butterworth-Heinemann; 1999.

[6] Flumerfelt RW, Glover CJ. Transport phenomena. Ullmann's chemical engineering and plant design—volume 1. Mathematics and physics in chemical engineering. Fundamentals of chemical engineering. Weinheim, Germany: Wiley-VCH; 2005.

[7] Eckhoff RK. Understanding dust explosions. The role of powder science and technology. Journal of Loss Prevention in the Process Industries 2009;22:105–16.

[8] Mercer DB, Amyotte PR, Dupuis DJ, Pegg MJ, Dahoe A, de Heij WBC, Zevenbergen JF, Scarlett B. The influence of injector design on the decay of pre-ignition turbulence in a spherical explosion chamber. Journal of Loss Prevention in the Process Industries 2001;14:269–82.

[9] Skjold T. Review of the DESC project. Journal of Loss Prevention in the Process Industries 2007;20:291–302.

[10] Dahoe AE, Cant RS, Scarlett B. On the decay of turbulence in the 20-liter explosion sphere. Flow, Turbulence and Combustion 2001;67:159–84.

[11] Abbasi T, Abbasi SA. Dust explosions—cases, causes, consequences, and control. Journal of Hazardous Materials 2007;140:7–44.

[12] Sanchirico R, Di Benedetto A, Garcia-Agreda A, Russo P. Study of the severity of hybrid mixture explosions and comparison to pure dust-air and vapour-air explosions. Journal of Loss Prevention in the Process Industries 2011;24:648–55.

[13] Zalosh R. Explosion venting data and modeling research project. Literature review Quincy, MA: The Fire Protection Research Foundation; 2008.

[14] Eckhoff RK. Dust explosions in the process industries, 3rd ed. Boston, MA: Gulf Professional Publishing/Elsevier; 2003.

[15] Klippel A, Scheid M, Krause U. Investigations into the influence of dustiness on dust explosions. Krakow, Poland: Proceedings of Ninth International Symposium on Hazards, Prevention, and Mitigation of Industrial Explosions; July 22–27, 2012.

[16] Kalejaiye O, Amyotte PR, Pegg MJ, Cashdollar KL. Effectiveness of dust dispersion in the 20-L Siwek chamber. Journal of Loss Prevention in the Process Industries 2010;23:46–59.

[17] Holbrow P, Andrews S, Lunn GA. Dust explosions in interconnected vented vessels. Journal of Loss Prevention in the Process Industries 1996;9:91–103.

[18] Kosinski P, Hoffmann AC. An investigation of the consequences of primary dust explosions in interconnected vessels. Journal of Hazardous Materials 2006;A137:752–61.

[19] Amyotte PR, Pegg MJ. Lycopodium dust explosions in a Hartmann bomb: effects of turbulence. Journal of Loss Prevention in the Process Industries 1989;2:87–94.

[20] Kauffman CW, Srinath SR, Tezok FI, Nicholls JA, Sichel M. Turbulent and accelerating dust flames. Proceedings of the 20th International Symposium on Combustion, Pittsburgh, PA: The Combustion Institute; 1985; pp. 1701–8.

Chapter 15

Myth No. 14 (Confinement): Venting Is the Only/Best Solution to the Dust Explosion Problem

We now complete the explosion pentagon with a look at some of the possible myths concerning *confinement*. This is the second key element in addition to mixing that marks the primary distinction between a *fire* and an *explosion*. The current chapter deals, in part, with the relief of confinement by means of explosion venting. The treatment is not, however, an in-depth review of venting technology (see, for example, Barton [1]), but rather an examination of where venting fits in the overall scheme of dust explosion risk reduction. The following two chapters cover other aspects beyond the more traditional levels of confinement provided by totally enclosed process vessels.

An explosion vent is a weak area in the wall of a building (e.g., a relief panel) or a dust-handling unit (e.g., a rupture disk) that is designed to open at an early stage in an explosion [1]. A mixture of combustion products and burning and unburned dust is typically expelled from the enclosure upon vent opening [1]. If venting achieves its desired goal, then overpressure is reduced to a value that does not exceed the enclosure strength. This process is shown by schematic in Figure 15.1 and by photograph in Figure 15.2.

Venting is arguably the most common approach taken to mitigate the overpressure developed during an industrial dust explosion—so much so that it may sometimes seem to be the *only* or *best* approach available to deal with the dust explosion problem. This is most definitely not to imply that venting is overrated with respect to its usefulness in explosion protection. Rather, the suggestion being made here is that venting—and other ways of preventing and mitigating explosions—should be viewed not as isolated techniques, but as components of a suite of risk reduction methodologies.

Critical to this way of thinking is the concept of inherently safer design—or more simply, inherent safety.

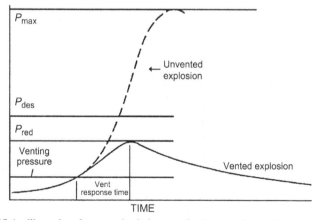

FIGURE 15.1 Illustration of pressure rise during vented and unvented explosions. P_{max} = maximum explosion pressure generated in the unvented enclosure; P_{des} = maximum pressure the protected enclosure can withstand; P_{red} = maximum pressure allowed to be generated in the enclosure during explosion; Venting pressure = pressure at which the vent opens (also known as P_{stat}) [2].

15.1 INHERENTLY SAFER DESIGN

Inherent safety is a proactive approach in which hazards are eliminated or lessened so as to reduce risk with decreased reliance on engineered (add-on) devices and procedural measures. The concepts of inherent safety and inherently safer design have been formalized over the past 35 or so years, beginning with the pioneering work of Professor Trevor Kletz (largely in response to the cyclohexane explosion at Flixborough, UK, in 1974 [4]). (See Section 11.3.) A number of principles or guidewords to facilitate inherent safety implementation in industry have been developed; the four basic principles identified in Table 15.1 have gained widespread acceptance.

Amyotte et al. [6] and Chapter 13 in Kletz and Amyotte [7] give numerous examples of application of the principles shown in Table 15.1 to the issue of dust explosion risk reduction. A recapitulation from Amyotte and Eckhoff [8]

FIGURE 15.2 Example of a vented dust explosion (corn flour at a dust concentration of 750 g/m^3 in a vessel of volume 2 m^3 with a vent area of 0.2 m^2 and central ignition) [3].

TABLE 15.1 Key Inherent Safety Principles [5]

Principle	Description
Minimization	Use smaller quantities of hazardous materials when the use of such materials cannot be avoided or eliminated. Perform a hazardous procedure as few times as possible when the procedure is unavoidable.
Substitution	Replace a substance with a less hazardous material or processing route with one that does not involve hazardous material. Replace a hazardous procedure with one that is less hazardous.
Moderation	Use hazardous materials in their least hazardous forms or identify processing options that involve less severe processing conditions.
Simplification	Design processes, processing equipment, and procedures to eliminate opportunities for errors by eliminating excessive use of add-on safety features and protective devices.

includes the following points, many of which have been presented in earlier chapters from a fundamental perspective:

- Minimization
  - Avoidance of the formation of combustible dust clouds. Because of the large quantities of particulate material present in powder handling equipment (which, as previously mentioned, is where most primary dust explosions occur), it can be difficult, however, to achieve operation at dust concentrations below the minimum explosible concentration [9].
  - Removal of dust deposits (avoidance of dust layers). Minimization of fuel loadings in this case is critical to the prevention of secondary dust explosions as discussed in Chapter 12.
- Substitution
  - Replacement of bucket elevators and other mechanical conveying systems with dense-phase pneumatic transport.
  - Substitution of process hardware with less hazardous materials of construction (e.g., avoiding unnecessary use of insulating materials [10]).
  - Use of mass flow silos and hoppers rather than funnel flow silos so as to avoid undesired particle segregation and uncontrolled dust cloud formation [11].
  - Alteration of a process route that involves handling an explosible powder (e.g., earlier introduction of an inert powder that is a component of the final product [12]).
  - Replacement of a combustible dust with one that is less hazardous. While this may be difficult to achieve in many cases, opportunities can arise when other factors such as cost motivate process change [13]. For example, Figure 15.3 illustrates that petroleum coke is a safer fuel than higher volatile-matter coal (from the perspective of rate of pressure rise). When used in a blended fuel as a partial replacement for pulverized coal in the feed to utility boilers, this inherent safety benefit of petroleum coke can manifest itself as shown in Figure 15.4.
- Moderation
  - Altering the composition of a dust by admixture of solid inertants [14,15]. (See Chapter 7; note that this inherent safety example is for inerting, not suppression.)
  - Increasing the dust particle size so as to decrease its reactivity [15,16]. (See Chapter 6.)
  - Avoiding the formation of hybrid mixtures of combustible dusts and flammable gases [15,16]. (See Section 5.4.)
  - Using powders in paste or slurry form. (Recall the discussion in Section 12.3 on the West Pharmaceutical Services incident.)
- Simplification
  - Employing the concept of error tolerance by designing process equipment robust enough to withstand process upsets and other undesired events

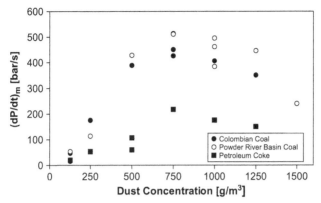

FIGURE 15.3 Influence of fuel type on rate of pressure rise of coal and petroleum coke fuels; tests were done using a Siwek 20-L chamber with 5-kJ ignition energy [13].

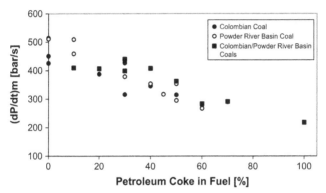

FIGURE 15.4 Reduction in rate of pressure rise achieved by using blended fuels of petroleum coke and two different coals as well as a mixture of the two coals (50 weight % of each); tests were done using a Siwek 20-L chamber with 5-kJ ignition energy and a dust concentration of 750 g/m³ [13].

(e.g., shock- or pressure-resistant design). An example here is the hammer-mill shown in Figure 3.9. (See Section 3.2.)

- Ensuring information on the hazardous properties of combustible dusts is clear and unambiguous (e.g., by means of thorough and complete material safety data sheets as discussed in Chapter 20).

15.2 HIERARCHY OF CONTROLS

The inherent safety principles given in Table 15.1 (and others [7]) work in conjunction with additional means of reducing risk—namely, passive and active engineered safety, and procedural safety—within a framework commonly known as the *hierarchy of controls* [17] (or as some call it, the *priority of controls* or the *safety decision hierarchy* [18]). Inherent safety, being the most effective approach to risk reduction, sits at the top of the hierarchy; it is followed in order of decreasing effectiveness by passive engineered safety devices (e.g., explosion relief vents), then active engineered safety devices (e.g., automatic suppression systems), and finally procedural safety measures (e.g., ignition source control by hot-work permitting).

Figure 15.5 shows a widely applicable approach to loss prevention that incorporates this preferred order of controls to deal with process hazards and the ensuing risk. Here, unit segregation is broken out separately from inherent safety; this feature can also be categorized under the *moderation* subprinciple of *limitation of effects*. The hierarchical arrangement in Figure 15.5 is similar to the layer of protection analysis (LOPA) concept in which inherently safer process design sits at the central core of the layers [19].

15.3 DUST EXPLOSION PREVENTION AND MITIGATION MEASURES

Drawing on the breakdown of dust explosion control measures given by Eckhoff [9] and their categorization according to Figure 15.5, Table 15.2 provides a hierarchical view of these measures. The presentation is similar to that given by Amyotte and Eckhoff [8] with the additional recognition of the procedural safety aspects of housekeeping. (See Section 4.3.)

As noted in several other instances in Table 15.2, there is often an overlap in the hierarchy of controls. This is consistent with the representation of the hierarchy categories in Figure 15.6 as a spectrum of options rather than as discrete entities having sharply defined boundaries [21]. Hendershot [21] further remarks that while people may disagree about the category in which a given approach falls, what really matters is whether the approach is effective from an engineering design viewpoint. I could not agree more.

15.4 REALITY

There are many ways to prevent the occurrence and mitigate the consequences of dust explosions, as summarized in Table 15.2 from a hierarchical perspective. Points to note in using Table 15.2 as a guide for dust explosion risk reduction measures include [8,21,22]

- Inherent safety is the most effective way to deal with a dust explosion hazard. To paraphrase Professor Trevor Kletz and the title of his seminal paper [23] on inherent safety: "What you don't have, can't explode."
- Inherent safety achieves its greatest impact when considered early in the design life cycle. Once a process unit is built to withstand only moderate pressure excursions, it may be too late to consider an explosion pressure-resistant design for that particular unit.
- Inherent safety is hazard-specific and should not be expected to eliminate all hazards. (Recall the discussion of the Kleen Energy explosion in Section 11.2.)

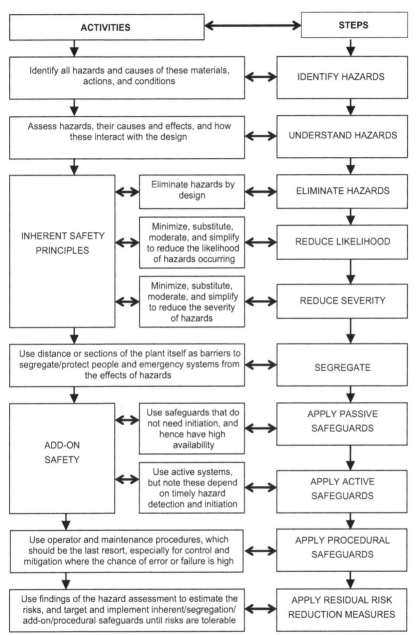

FIGURE 15.5 A systematic approach to loss prevention incorporating the hierarchy of controls (adapted from Kletz and Amyotte [7] with the modification that the inherent safety principles of minimization, substitution, moderation, and simplification can all be effective for both prevention and mitigation purposes [20]).

TABLE 15.2 A Hierarchical View of Various Means of Preventing and Mitigating Dust Explosions

Explosion prevention		Explosion mitigation
Preventing explosible dust clouds	Preventing ignition sources	
Process design to prevent undesired generation of dust clouds and particle size reduction and segregation *Inherent Safety— Minimization, Substitution, Moderation, Simplification*	Smoldering combustion in dust, dust fires *Procedural Safety—may also involve aspects of Inherent Safety or Engineered Safety*	Good housekeeping (dust removal/cleaning) Mitigation with respect to secondary dust explosions; prevention with respect to primary dust explosions *Inherent Safety— Minimization Procedural Safety*
Keeping dust concentration outside explosible range *Inherent Safety— Minimization*	Other types of open flames (e.g., hot work) *Procedural Safety—may also involve aspects of Inherent Safety or Engineered Safety*	Explosion-pressure resistant construction *Inherent Safety— Simplification*
Inerting of dust cloud by adding inert dust *Inherent Safety— Moderation*	Hot surfaces (electrically or mechanically heated) *Procedural Safety—may also involve aspects of Inherent Safety or Engineered Safety*	Explosion isolation (sectioning) *Inherent Safety—Moderation (e.g., unit segregation, product choke) if not using mechanical devices. If mechanical devices are used to isolate plant sections, classification would be Engineered Safety—Passive in the case of physical barriers, or Engineered Safety—Active in the case of isolation valves.*
Intrinsic inerting of dust cloud by combustion gases *Engineered Safety—Active*	Heat from mechanical impact (metal sparks and hot-spots) *Procedural Safety—may also involve aspects of Inherent Safety or Engineered Safety*	Explosion venting *Engineered Safety—Passive*
Inerting of dust cloud by N_2, CO_2, and rare gases *Engineered Safety—Active*	Electric sparks and arcs and electrostatic discharges *Procedural Safety—may also involve aspects of Inherent Safety or Engineered Safety*	Automatic explosion suppression *Engineered Safety—Active*
		Partial inerting of dust cloud by inert gas *Engineered Safety—Active*

(adapted from Amyotte and Eckhoff [8])

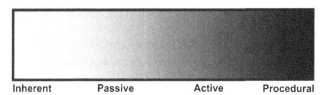

| Inherent | Passive | Active | Procedural |

FIGURE 15.6   Process safety strategies represented as a spectrum of options ranging from inherent through to procedural [21].

- Some inherent safety options may not be feasible in a given application. For example, a chemically inert dust cannot likely be substituted for an explosible dust if the latter is the actual desired product. But if the relevant inherent safety questions are not asked, then potential process or product alternatives cannot be explored.
- Inherent safety is not a stand-alone concept. It works through a hierarchical arrangement in concert with engineered (passive and active) and procedural safety to reduce risk.
- The hierarchy of controls does not invalidate the usefulness of engineered and procedural safety measures. Quite the opposite—the hierarchy of controls recognizes the importance of engineered and procedural safety by highlighting the need for careful examination of the reliability of both mechanical devices and human actions. These considerations must be incorporated into the dust explosion risk assessment process.

The preceding points are well illustrated by a case study concerning the handling of dry additive at a polyethylene production facility [15]. Key inherent safety principles highlighted are minimization, substitution, and moderation. Additionally, the example demonstrates the economic trade-offs and interrelationship among inherent, engineered, and procedural safety measures.

In the late 1970s, dry additive was received at the plant in heavy 50-kg containers. Operators were required to scoop additive into a feeder that supplied the additive to a pre-blender to ultimately mix additive with polyethylene resin to achieve certain resin properties. This activity caused concern over back strain as well as the need to wear respirators to control exposure to the additive dust. In the late 1980s and early 1990s, an effort was made to improve working conditions and efficiency. A capital project was proposed to pneumatically convey additive to the feeder. The additive is granular and, with a minimum ignition energy of less than 10 mJ, is also ignition sensitive.

A suggestion was made to consider the process with nitrogen as the conveying medium. A number of issues associated with this option soon materialized:

- Higher operating costs using nitrogen (recycling nitrogen posed technical challenges; additionally, this would have been the first attempt at the site in conveying solids using nitrogen medium)
- The need to monitor and control oxygen content so as not to exceed the limiting oxygen concentration

- A requirement for greater operator attention if a manual monitoring approach was adopted
- Prohibitively high project costs for automatic monitoring and alarms
- Possible asphyxiation should nitrogen vent inside a building.

Because of these concerns, the nitrogen conveying option was not considered feasible. Operations returned to manual handling of the dry additive, but this time using smaller containers (25 kg) to deal with back strain concerns. The site was also continuously working with the additive supplier to identify better approaches (e.g., to reduce manual handling). This gave rise to experimenting with a pellet-like version of the same additive in a pneumatic conveying system. This would, in theory, also remove the concern over the presence of a combustible dust cloud having a low ignition energy inside an air conveying system. However, the mechanical energy of the pneumatic conveying system easily broke down the pellet-like additive into a very fine powder, causing extensive dust buildup and resulting in operating problems due to plugging. The dust explosion concern was also reintroduced.

In the late 1990s, the site installed its current system, which involves the use of supersacks and tote tanks. Supersacks from the supplier are emptied by gravity into tote tanks at grade level. These totes are then taken by elevator to the appropriate floor where they are placed on top of a feed pipe system that connects to the additive feeder. The feed pipe system is filled with additive by opening a slide valve located below the tote. The slide valve is opened slowly to avoid disturbing the sensitive operation of the feeder. This also helps with minimizing the formation of dust clouds inside the pipe system as it is being filled. This current option has eliminated the following safety concerns:

- Repetitive manual handling by operators and associated back strain
- The need to wear respirators to avoid inhalation of dust
- Excessive static charging and combustible dust cloud formation that would have been associated with pneumatic conveying, and the possible ignition of the ignition-sensitive dust inside an air conveying system.

15.5 WHAT DO *YOU* THINK?

Process Safety—the prevention and mitigation of process-related injuries and damage arising from process incidents (fire, explosion, and toxic release).

A recurrent theme in this book is, I hope, now starting to become more apparent and prominent: Dust explosions are process incidents, and process incidents can be effectively prevented and mitigated only by the rigorous application of process safety principles. We have just seen one of the most important process safety principles—inherent safety. The suggested exercises in this section encourage a close relationship between inherently safer design and dust explosion risk reduction.

The first exercise is to conduct an examination of your facility for opportunities to apply the key inherent safety principles of minimization, substitution, moderation, and simplification to the dust explosion problem. You may find it helpful to use or develop a checklist tailored to this purpose. Kletz and Amyotte [7] give several ideas for general inherent safety checklist questions; they might be useful in their original form or modified to specifically address combustible dusts. For example [7]:

- *Minimize*—Are all hazardous materials removed or properly disposed of when they are no longer needed or not needed in the next x days?
- *Substitute*—Can a less toxic, less flammable, or less reactive material be substituted for use?
- *Moderate*—Are all hazardous gases, liquids, and solids stored as far away as possible to eliminate disruption to people, property, production, and environment in the event of an incident (avoiding knock-on effects)?
- *Simplify*—Are all manuals, guides, and instructional materials clear and easy to understand, especially those that are used in an emergency situation?

The second exercise relates to the training of coworkers on the relationship between inherent safety and dust explosions. (The discussion here parallels that given for training on inherent safety and hydrogen safety by Kletz and Amyotte [7] and Rigas and Amyotte [24], respectively.) In a typical training protocol, the second step after conducting a needs analysis is the setting of learning (instructional) objectives. Excellent advice is given by Felder and Brent [25] on this point.

Felder and Brent [25] comment that when one is setting such objectives, four leading verbs should be avoided: *know, learn, appreciate,* and *understand.* Thus, while it would be desirable for plant employees to know, learn, appreciate, and understand the principles of inherent safety as applied to dust explosion prevention and mitigation, these are not valid instructional objectives because it is not possible to directly see whether they have been done. It is necessary to consider what trainees should be asked to *do* to demonstrate their knowledge, learning, appreciation, and understanding of inherent safety as it relates to dust explosions. These activities should then become the instructional objectives [25].

Felder and Brent [25] further describe the concept of using action verbs to set instructional objectives. Using their breakdown according to the classification scheme known as *Bloom's Taxonomy of Educational Objectives*, appropriate instructional objectives for the relationship between inherent safety and dust explosions would be (with the action verb underlined):

- *Knowledge*—State the preferred order of safety measures in the hierarchy of controls and give an example in each hierarchy category for dust explosion prevention or mitigation.
- *Comprehension*—Explain in your own words the concept of moderation as applied to combustible dusts.

- *Application*—<u>Calculate</u> the airborne dust concentrations resulting from the dispersion of several dusts of different bulk densities for a fixed dust layer thickness and enclosure height. The purpose is to ascertain the most significant hazard scenario according to the principle of minimization.
- *Analysis*—<u>Classify</u> the causation factors for a dust explosion incident using the principles of inherent safety as a guide for best practice.
- *Synthesis*—<u>Formulate</u> an inherently safer alternative to a given design for combustible dust conveyance.
- *Evaluation*—<u>Determine</u> which of several dust explosion mitigation schemes is inherently safer and explain your reasoning.

You could now try writing your own instructional objective for each of the six levels in the taxonomy. As a guide, here are some additional action verbs for each level [25]: (i) Knowledge—identify, summarize; (ii) Comprehension—describe, interpret; (iii) Application—apply, solve; (iv) Analysis—derive, explain; (v) Synthesis—design, create; and (vi) Evaluation—optimize, select. Recall that the overall objective is to relate the principles of inherent safety to the topic of dust explosion prevention and mitigation.

> Note: Although Bloom's Taxonomy has been revised in recent years, the original form as indicated in the preceding text has been retained here for consistency with the work of Felder and Brent [25]. Much good would result from the setting of learning/instructional objectives according to either the original or revised version of the taxonomy.

The final exercise asks you to practice your drawing skills and develop some simple graphics to illustrate the four basic inherent safety principles (Table 15.1) within the context of combustible dusts. First, consider Figure 15.7 meant to illustrate the essence of the general simplification principle. I would say this is a good example of simple pictures being more powerful than words for conveying ideas, as quoted in Kletz and Amyotte [7].

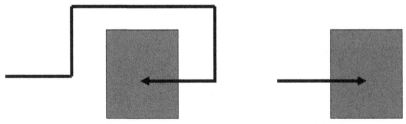

FIGURE 15.7 A graphical representation of the inherent safety principle of simplification (adapted from Kletz and Amyotte [7]).

I would also say that Figure 15.7 more than adequately summarizes the lessons of the following story [7] (pp. 158–159):

An air-weary traveler arrived at an airport and walked through the sliding glass doors as he had done at scores of similar airports in the past. Expecting the second set of glass doors in front of him to slide open in the usual fashion, he merrily proceeded forward only to be brought up short when no opening appeared. He then realized that he was facing a glass wall, not more sliding doors. There on the glass partition in front of him was a thin line of yellow caution tape with small arrows directing him to turn to his left. He did so and there indeed was the second set of glass doors which opened so he could walk through them, turn to his right and then finally approach the check-in counter.

Intrigued, and a little embarrassed, our intrepid traveler stood back and watched to see how others would cope with this seemingly unnecessary complication. Person after person fared no better, and some even worse, in navigating the glass maze.

So why did the airport management decide to challenge its customers in this manner? The apparent reason is that had the two sets of sliding doors been arranged in the normal straight-through design, a blast of cold air would have reached the ticket agents at their stations during the winter months. One can surmise that it was the custom at this particular airport to locate the agents close to the doors for the convenience of all concerned. But was this really an appropriate specification for a location with a somewhat brisk winter climate?

The solution to this issue lies, of course, in the building design which once implemented, affords little opportunity for the use of revolving doors, automatic interlocks and the like (or even better, relocation of the check-in counters). But even when procedural safety fails as in the case of people not paying attention to directional signage, other solutions exist. On another visit to the same airport a few months later, the same traveler was prepared to follow the arrows and avoid a collision with the glass wall. He need not have worried; there in front of the glass wall was an enormous orange traffic cone [a passive barrier]—the largest he had ever seen!

Now to return to the third exercise of drawing some simple pictures to illustrate the four basic inherent safety principles and their relationship to dust explosions. I am not much of an artist but some initial thoughts do come to mind. How about *moderation* as it relates to changes in particle size (smaller to larger)? Or *minimization* as it might be expressed in terms of "no dust layers permitted"?

REFERENCES

[1] Barton J, editor. Dust explosion prevention and protection. A practical guide. Rugby, UK: Institution of Chemical Engineers; 2002.
[2] Pekalski AA, Zevenbergen JF, Lemkowitz SM, Pasman HJ. A review of explosion prevention and protection systems suitable as ultimate layer of protection in chemical process installations. Process Safety and Environmental Protection 2005;83:1–7.

[3] Holbrow P. Dust explosion venting of small vessels and flameless venting. Process Safety and Environmental Protection, 2012; http://dx.doi.org/10.1016/j.psep.2012.05.003; (last accessed March 24, 2013).

[4] Kletz T. What went wrong? Case histories of process plant disasters and how they could have been avoided, 5th ed. Oxford, UK: Gulf Professional Publishing; 2009.

[5] Amyotte PR. Dust explosions happen because we believe in unicorns. Keynote Lecture. College Station, TX: Proceedings of 13th Annual Symposium, Mary Kay O'Connor Process Safety Center, Texas A&M University; October 26–28, 2010, pp. 3–30.

[6] Amyotte PR, Pegg MJ, Khan FI. Application of inherent safety principles to dust explosion prevention and mitigation. Process Safety and Environmental Protection 2009;87:35–9.

[7] Kletz T, Amyotte P. Process plants. A handbook for inherently safer design, 2nd ed. Boca Raton, FL: CRC Press, Taylor & Francis Group; 2010.

[8] Amyotte PR, Eckhoff RK. Dust explosion causation, prevention and mitigation: an overview. Journal of Chemical Health & Safety 2010;17:15–28.

[9] Eckhoff RK. Dust explosion prevention and mitigation. Status and developments in basic knowledge and in practical application. International Journal of Chemical Engineering 2009 2009; Article ID 569825, 12 pp.

[10] Kong D. Analysis of a dust explosion caused by several design errors. Process Safety Progress 2006;25:58–63.

[11] Eckhoff RK. Understanding dust explosions. The role of powder science and technology. Journal of Loss Prevention in the Process Industries 2009;22:105–16.

[12] Mintz KJ, Bray MJ, Zuliani DJ, Amyotte PR, Pegg MJ. Inerting of fine metallic powders. Journal of Loss Prevention in the Process Industries 1996;9:77–80.

[13] Amyotte PR, Basu A, Khan FI. Reduction of dust explosion hazard by fuel substitution in power plants. Process Safety and Environmental Protection 2003;81:457–62.

[14] Amyotte PR. Solid inertants and their use in dust explosion prevention and mitigation. Journal of Loss Prevention in the Process Industries 2006;19:161–73.

[15] Amyotte PR, Pegg MJ, Khan FI, Nifuku M, Yingxin T. Moderation of dust explosions. Journal of Loss Prevention in the Process Industries 2007;20:675–87.

[16] Amyotte PR, Cloney CT, Khan FI, Ripley RC. Dust explosion risk moderation for flocculent dusts. Journal of Loss Prevention in the Process Industries 2012;25:862–9.

[17] Hopkins A. Safety, culture and risk. The organizational causes of disasters. Sydney, Australia: CCH Australia Limited; 2005.

[18] Manuele FA. Risk assessment & hierarchies of control. Professional Safety 2005;50:33–9.

[19] Dowell AM. Layer of protection analysis and inherently safer processes. Process Safety Progress 1999;18:214–20.

[20] Amyotte PR, MacDonald DK, Khan FI. An analysis of CSB investigation reports concerning the hierarchy of controls. Process Safety Progress 2011;30:261–5.

[21] Hendershot DC. A summary of inherently safer technology. Process Safety Progress 2010;29:389–92.

[22] Amyotte PR, Khan FI, Dastidar AG. Reduce dust explosions the inherently safer way. Chemical Engineering Progress 2003;99(10):36–43.

[23] Kletz TA. What you don't have, can't leak. Chemistry and Industry 1978;6:287–92.

[24] Rigas F, Amyotte P. Hydrogen safety. Boca Raton, FL: CRC Press, Taylor & Francis Group; 2012.

[25] Felder RM, Brent R. Objectively speaking. Chemical Engineering Education 1997;31:178–9.

Chapter 16

Myth No. 15 (Confinement): Total Confinement Is Required to Have an Explosion

Hertzberg and Cashdollar [1] (p. 5) define an *explosion* in practical terms as "a gas-dynamic phenomenon characterized by such a rapid increase in system pressure that destructive forces are generated." Using the familiar ideal gas law, they further explain how the rapid oxidation of dust particles dispersed in a cloud leads to a similarly rapid increase in temperature and an ensuing pressure increase [1]. Noting that the molar amount of gas (n) present in a system usually does not vary significantly during the combustion process, pressure (P) must trend as temperature (T) does in a fixed volume (V) given that pressure and temperature are on opposite sides of the ideal gas law equation. In other words, with the only other term in the equation being the universal gas constant (R), pressure changes must be proportional to temperature changes for the equal sign to hold true.

The preceding analysis is admittedly simple and certainly not exact at all times. Recall the discussion in Section 7.3 on the use of sodium azide, NaN_3, as a gas generant to inflate automobile airbags. In cases such as this, the molar amount of gas is obviously not a near-constant. Yet the ideal gas law does help to illustrate the role of confinement via fixed volumes in facilitating overpressure development. One needs to be careful, however, not to give the impression that complete or total confinement is an absolute prerequisite for a dust explosion. This same notion can also be inadvertently conveyed by the graphical representation of the explosion pentagon showing a closed polygon. (See Figure 1.3.)

Thus, there is more to the story of confinement than dust explosions occurring in totally enclosed environments—whether the system boundaries are rigid or are able to expand in either a desired or undesired manner. Hertzberg and Cashdollar [1] accordingly extend their discussion of confinement effects to account for both partially confined and unconfined explosions. In the current chapter, we briefly examine how a dust explosion could arise, or the consequences be intensified, under such circumstances. The following chapter completes our discussion of confinement with a look at scenarios that do not involve the typical boundaries presented by process vessels, buildings, and other permanent structures.

16.1 DEGREE OF CONFINEMENT

Abbasi et al. [2] propose the classification scheme shown in Figure 16.1 for explosions occurring in the chemical process industries (CPI). Here, we see dust explosions (middle of the bottom row) identified as events that can occur in an unconfined environment but are more likely in a confined or partially confined space. Abbasi et al. [2] further classify dust explosions into three categories according to whether they arise in (i) unconfined spaces with a sufficiently high blockage ratio due to flow obstruction resulting in high flame speeds (the subject of Section 17.1); (ii) semi-confined spaces such as empty building areas or corridors found, for example, in coal mines (as further described in the current section); or (iii) confined or vented spaces such as process units for mixing, drying, separating, etc. (the subject of Section 16.2).

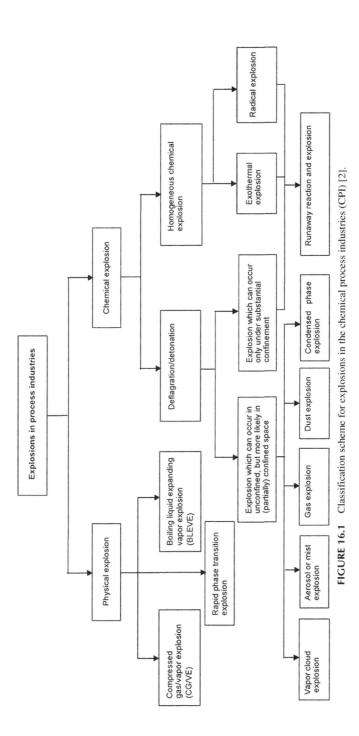

FIGURE 16.1   Classification scheme for explosions in the chemical process industries (CPI) [2].

The preceding points are generally consistent with Hoyle's remarks [3] on how confinement influences the combustible dust hazard, with ignition leading to (i) a flash fire with the potential for burn injuries and secondary fires (no confinement); (ii) a fireball with limited pressure rise inside, and flame propagation outside, the enclosure (partial confinement); and (iii) full overpressure development with subsequent destruction of buildings and process units unless these are robustly designed (complete confinement).

Further on the matter of dust explosions in unconfined spaces, these are expected to be rare occurrences because of the requirement for pressure buildup in the dust cloud at a rate exceeding pressure dissipation at the cloud edge [4]. This would necessitate rapid combustion reactions such as could be brought about by the aforementioned flow obstruction scenario combined with a high level of dust reactivity. (See also Section 17.1.) Flash fires are therefore prevalent in the case of low levels of confinement [3] and in incidents like those occurring with iron dust at the Hoeganaes facility in Gallatin, TN [5]. (Recall the suggested exercise given in Section 1.4. and see also Sections 12.1 and 16.4.)

With respect to partial confinement (item [ii] in the preceding listings from Abbasi et al. [2] and Hoyle [3]), explosions in underground mine workings represent an illustrative example [6–8]. Explosion development and flame propagation in this case can be approximated by considering a mine gallery to be a corridor with one end open and ignition occurring at the other closed end [1]. (Examples also exist in the literature of partially confined experimental apparatus used to study dust flame propagation [9].)

The piston-like effect of the burned gases expanding behind the flamefront pushes the unburned fuel/air mixture toward the open end and results in the generation of post-ignition turbulence in the unburned mixture. The advancing flame front then accelerates as it encounters the turbulent flow field with the end result being a self-accelerating feedback mechanism between flame speed and turbulence level in the unreacted flow field [1]. A look back at Figure 3.2 gives a sense of the severe consequences to which this process can lead.

16.2 EXPLOSION RELIEF VENTING

As mentioned in Chapter 15, venting is the most common approach taken to mitigate dust explosion overpressures, the purpose being to essentially relieve the confinement criterion of the explosion pentagon. In doing so, however, it is critical to maintain awareness of some basic requirements for effective venting. First and foremost, it should be well understood that an undersized vent can lead to a situation in which the design pressure of the protected enclosure is exceeded. (See P_{des} in Figure 15.1). Second, vents must be designed to open freely and without obstruction when called into service [10]; otherwise the degree of confinement will remain too high (i.e., too "close to complete") for sufficient overpressure relief to occur.

With flame propagation through vent openings as shown in Figure 15.2, measures must also be taken to protect plant personnel [10]. One approach is the use of vent ducts to direct the burned, burning, and unburned dust ejected out of a vent to a safer location [11]. Care is needed here to ensure that the presence of the vent duct is taken into account with respect to obstruction of the venting process itself [11,12]. Referring again to Figure 15.1, the reduced explosion pressure, P_{red}, determined for a stand-alone vented vessel may, in fact, be increased to a value above the vessel design pressure, P_{des}, by the addition of a vent duct [11].

As an alternative to vent ducts, and to avoid the ejection of dust and flames through vent openings, a technique known as *flameless venting* has become increasingly popular. A flame arresting device typically composed of mesh layers with quenching channels is fitted in conjunction with the vent to limit the escape of flame and burning dust [13]. Figure 16.2 shows such a device fitted to the same test rig as in Figure 15.2. Here, we see the effectiveness of the flameless venting device in eliminating flame propagation through the vent and permitting the passage of only smoke, water vapor, and dust (i.e., combustion products and unconsumed fuel) [14].

Cautionary notes for the use of flame arrestors in conjunction with explosion relief vents are twofold. First, the possibility of reduced venting efficiency must be considered. Holbrow [14] determined that the performance of flameless venting devices is likely a function of dust characteristics other than K_{St}, the size-normalized maximum rate of pressure rise. When such a device was used, no reduction in venting efficiency (as measured by an increase in reduced pressure, P_{red}) was observed for wheat flour with a K_{St} of 138 bar·m/s; on the other hand, corn flour with a K_{St} of 147 bar·m/s showed a marked increase in P_{red} (i.e., a decrease in venting efficiency).

FIGURE 16.2   Example of a vented dust explosion with a flameless venting device installed (corn flour at a dust concentration of 750 g/m^3 in a vessel of volume 2 m^3 with a vent area of 0.2 m^2 and central ignition) [14].

Second, manufacturer literature provides clear warning that flameless venting into an enclosed environment will generate pressure within that enclosure (e.g., a building if venting is indoors) [13]. Possible overpressures should therefore be compared to the enclosure design pressure; venting of the enclosure itself may be advisable [13].

Thus, while venting is an intrinsically simple concept, subject matter knowledge is essential for its use to be effective. Data from standardized explosion testing are required—as is the calculation of vent areas for specific protection scenarios according to applicable codes, standards, and best-practice methodologies. I have witnessed firsthand the effects on industrial plant of improperly designed and located explosion vents. No matter how well intentioned the motivation of the plant engineers undertaking these efforts, the results were not indicative of practice within a defined area of expertise.

16.3 REALITY

The development of destructive overpressures is most often associated with complete confinement of a rapidly combusting dust/air mixture in an explosion scenario. A completely unconfined dust/air dispersion is most often associated with a flash fire in which the primary concern is the thermal flux impinging on plant personnel and equipment.

Between these extremes lies the realm of partial confinement. Dust explosions in these environments can still generate significant overpressure and flame travel in the flow direction that lessens the degree of confinement. The partial confinement may be a permanent aspect of the workplace (such as long corridors), or it may be brought about by the explosion mitigation measure known as venting.

In the latter case, provision must be made for the potential of counteracting the very act of confinement relief by the addition of features such as vent ducts and flameless venting devices. Venting is an extremely useful and efficient dust explosion risk reduction measure—but only if provision for its use is undertaken by qualified personnel having the requisite expertise.

16.4 WHAT DO YOU THINK?

For the application example in this chapter, we return to Chapter 1 and the suggested exercise given in Section 1.4 for the 2011 Hoeganaes incidents investigated by the U.S. Chemical Safety Board [5]. The intention now is to focus on the three separate incidents and attempt to gain further understanding of their origin by examining the levels of confinement and the materials involved in each.

The resource material for this exercise is readily available on the CSB website (www.csb.gov): (i) the investigation report/case study itself [5], and

(ii) the accompanying video [15]. Although I highly recommend watching the entire 14-minute video [15], even watching the first 2 minutes will be beneficial for the purposes of the following exercise.

Consider the three incidents:

- Iron dust flash fire on January 31, 2011
- Iron dust flash fire on March 29, 2011
- Hydrogen explosion and iron dust flash fire on May 27, 2011.

For each incident, what role did the degree of confinement play in determining whether the result of ignition of the fuel/oxidant mixture was a flash fire or an explosion? Note that there are four cases to consider: the three instances of iron dust flash fire occurrence (January 31, March 29, and May 27, 2011) as well as the hydrogen explosion on May 27, 2011.

What role did the material reactivity (or explosibility) play in each of the four cases with respect to ignition of the fuel/oxidant mixture resulting in a flash fire or in an explosion? Look in the case study [5] for data on maximum explosion pressure, size-normalized maximum rate of pressure rise, and minimum ignition energy of the iron dust present in the facility. You will also find information there on the minimum ignition energy of hydrogen. Maximum explosion pressure and size-normalized maximum rate of pressure rise values of 8.2 bar(g) and 503 bar·m/s, respectively, can be used for hydrogen [16].

You will notice that the case study [5] contains some details on comparison of iron dust explosibility data determined in 20-L and 1-m^3 chambers. The discussion is mostly (although not completely) peripheral to the exercise at hand in this section. It does serve, however, as an early introduction to the topic of *overdriving* to be covered in Chapter 19.

REFERENCES

[1] Hertzberg M, Cashdollar KL. Introduction to dust explosions. In: Cashdollar KL, Hertzberg M, editors. Industrial dust explosions. ASTM Special Technical Publication 958. Philadelphia, PA: American Society for Testing and Materials; 1987, pp. 5–32.

[2] Abbasi T, Pasman HJ, Abbasi SA. A scheme for classification of explosions in the chemical process industry. Journal of Hazardous Materials 2010;174:270–80.

[3] Hoyle M. Dust explosions. Risk to people, plant and reputation 2012; Process Safety Group, AstraZeneca. (PowerPoint presentation available at: http://www.process-safety-lab.com/UploadFiles/20111121184231418.pdf; last accessed November 18, 2012).

[4] Abbasi T, Abbasi SA. Dust explosions—cases, causes, consequences, and control. Journal of Hazardous Materials 2007;140:7–44.

[5] CSB. Case study—Hoeganaes Corporation: Gallatin, TN—metal dust flash fires and hydrogen explosion. Report No. 2011-4-I-TN. Washington, DC: U.S. Chemical Safety and Hazard Investigation Board; 2011.

[6] Cashdollar KL. Coal dust explosibility. Journal of Loss Prevention in the Process Industries 1996;9:65–76.

[7] Cashdollar KL, Sapko MJ, Weiss ES, Hertzberg M. Laboratory and mine dust explosion research at the Bureau of Mines. In: Cashdollar KL, Hertzberg M, editors. Industrial dust explosions. ASTM Special Technical Publication 958. Philadelphia, PA: American Society for Testing and Materials; 1987, pp. 107–23.

[8] Michelis J, Margenburg B, Muller G, Kleine W. Investigations into the buildup and development conditions of coal dust explosions in a 700-m underground gallery. In: Cashdollar KL, Hertzberg M, editors. Industrial dust explosions. ASTM Special Technical Publication 958. Philadelphia, PA: American Society for Testing and Materials; 1987, pp. 124–37.

[9] Gao W, Dobashi R, Mogi T, Sun J, Shen X. Effects of particle materials on flame propagation behaviour during organic dust explosions in a half-closed chamber. Krakow, Poland: Proceedings of Ninth International Symposium on Hazards, Prevention, and Mitigation of Industrial Explosions; July 22–27, 2012.

[10] Agarwal A. Dust explosions: prevention and protection. Chemical Engineering November 2012:26–30.

[11] Barton J, editor. Dust explosion prevention and protection. A practical guide. Rugby, UK: Institution of Chemical Engineers; 2002.

[12] Bartknecht W. Dust explosions. Course, prevention, protection. Berlin: Springer-Verlag; 1989.

[13] Fike. Explosion venting. Product guide. Blue Springs, MO: Fike Corporation; 2009.

[14] Holbrow P. Dust explosion venting of small vessels and flameless venting. Process Safety and Environmental Protection 2012, http://dx.doi.org/10.1016/j.psep.2012.05.003; (last accessed March 24, 2013).

[15] CSB. Iron in the fire (video). Washington, DC: U.S. Chemical Safety and Hazard Investigation Board; 2012. (Available at: http://www.csb.gov/investigations/detail.aspx?SID=100&Type=2&pg =1&F_All=y; last accessed November 19, 2012).

[16] Rigas F, Amyotte P. Hydrogen safety. Boca Raton, FL: CRC Press, Taylor & Francis Group; 2012.

Chapter 17

Myth No. 16 (Confinement): Confinement Means Four Walls, a Roof, and a Floor

The visual image created by the term *complete confinement* is typically one of an enclosure—perhaps a sphere or maybe a four-sided container with a top and bottom. The discussion of *partial confinement* in Section 16.1 relied on similar geometrical constructs such as a corridor, which can be approximated as an open-ended cylindrical tube. Common to all these confinement scenarios is the presence of physical boundaries. But it is not only permanent walls with spherical or rectangular dimensions that enhance dust flame propagation and overpressure development.

In this chapter we first examine a special category of post-ignition turbulence generation brought about by congestion and obstacles in the path of an advancing dust flame. The discussion on this point is a follow-up from that previously given for turbulence generated pre-ignition (Section 14.1) and also post-ignition (Section 16.1). The chapter concludes with a brief look at the issue of temporary structures leading to the unintentional consequence of increasing the level of confinement.

17.1 CONGESTION AND OBSTACLE-GENERATED TURBULENCE

Using the terminology of Crowl [1], a dust explosion would be called a *chemical* explosion (as opposed to a *physical* explosion) because it involves a chemical reaction—in this case, a *propagating* combustion reaction that is transmitted spatially through the reaction mass [1]. Propagating reactions can then be classified as either *deflagrations* or *detonations*, the distinguishing feature being the propagation (flame) speed of the reaction (flame) front through the unreacted mass (unburned dust cloud) [1].

In a deflagration, the reaction front travels at subsonic speed; in a detonation, the reaction front moves at sonic or supersonic speed (relative to the speed of sound in the unreacted medium) [1]. Thus, a deflagration is characterized by the flamefront trailing behind the blast or shock wave produced by the explosion; detonations involve a coupling of the reaction front and shock wave and can lead to overpressures significantly higher than those experienced with deflagrations. Eckhoff [2] describes detonation as an extreme mode of flame propagation through a dust cloud in which the deflagration mode of heat transfer from burning to unburned dust by molecular or turbulent diffusion is replaced with direct ignition of unburned dust by shock compression.

Detailed treatments of detonation theory and deflagration-to-detonation transition (or DDT) are beyond the scope of the current book. In essence, though, the discussion at the end of Section 16.1 from Hertzberg and Cashdollar [3] provides a good entry point for a qualitative understanding of these features. Recall that we left this examination of post-ignition turbulence generation in a corridor-like enclosure at the point where a self-accelerating feedback mechanism had been established between flame speed and turbulence level in the unreacted flow. With a sufficient run-up distance [4], wide flow channels [3], and additional turbulence created by wall surface roughness or, as we shall soon see, obstacles in the flow path, the result may be transition from a deflagration to a detonation.

Fortunately, dust detonations are not common in industry, and the vast majority of industrial dust explosions occur as deflagrations [5,6]. Detonations of combustible dust have, however, been shown to occur under laboratory conditions that optimize the geometry and flow criteria given in the preceding paragraph [7,8]. Additional dust properties that would increase the likelihood of a detonation include a fine particle size distribution, low moisture content, and ease of dispersal and entrainment [7]. Any industrial scenario involving these optimal geometry, flow, and dust property conditions would necessarily have a heightened likelihood of detonation. Although uncommon, industrial dust detonations are not unknown; possible candidates include long coal mine galleries [8], and long, wide lines as may be found in pulverized fuel power plants [5].

We saw previously in Section 16.1 how the severity of a dust explosion, although not prone to detonation, can be magnified by post-ignition (explosion-generated) turbulence. In addition to enhanced flame speeds caused by the piston-like action of combustion product expansion at the closed end of a tube or corridor, similar effects can arise from flow restriction and compression, or congestion in the downstream flow field.

The phenomenon known as *pressure-piling* can lead to substantial increases in explosion overpressure in an interconnected plant (i.e., interconnected process vessels) [9–11]. Lunn et al. [9] give an excellent description of the origin of the increased overpressure caused by an increased rate of dust combustion due to turbulence generated as the explosion passes from the vessel of origin through the interconnecting pipe. (This is similar to the effect of vent ducts as described in Section 16.2.) With the resulting compression of the unburned dust cloud in the second vessel, and ignition of this pre-compressed cloud by the flame entering the vessel, the explosion occurs in the second vessel at an initial pressure above ambient [9]. The net result is an increase in explosion overpressure over that expected in a single vessel; this effect is most pronounced in the case of explosion transmission from a large to a small vessel [9].

Obstacles in the path of an advancing dust flame can also create enough congestion or blockage to generate significant post-ignition turbulence—the result of which is higher flame speeds (i.e., flame acceleration) and more rapid pressure development. Figure 17.1 shows schematically the arrangement of obstacles in the laboratory-scale, closed-tube experimentation conducted by Zhou et al. [12] The influence of various obstacle geometries on the maximum rate of pressure rise during their testing of methane/coal dust hybrid mixtures is displayed by Figure 17.2. It can be seen that although the blockage ratio (ratio of obstacle area to tube cross-sectional area) was held constant at 40% in these tests, the more angular obstacles (e.g., square compared to circular) had a greater impact on maximum rate of pressure rise.

(a)   **(b)**   **(c)**

Hollow square   Torus   Semi-circle

FIGURE 17.1 Illustration (not to scale) of obstacle arrangement in coal dust/methane/air explosion studies (a - hollow square; b - torus; c - semi-circle) [12].

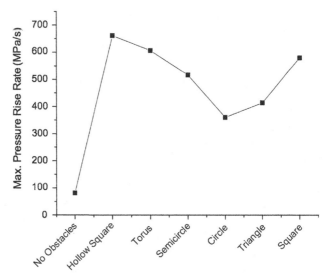

FIGURE 17.2 Influence of obstacle type on the maximum rate of pressure rise during closed-tube coal dust/methane/air explosions [12]. Tests were conducted with a methane concentration of 6.5 volume % and a 48-μm mean diameter coal dust concentration of 500 g/m^3.

Moen et al. [13] provide an explanation for the phenomenon of obstacle-induced flame acceleration based on photographic evidence from gas-phase deflagrations [8]. They have shown that an obstacle creates a wake pattern in the flow of unburned gases ahead of the flame. When the flame enters the turbulent flow region established by the obstacle, it becomes stretched and folded as the hot combustion zone is distorted and entrained into the wake by the action of large-scale eddies. An enhanced rate of burning and a higher burning velocity result from the increase in flame surface area. Small-scale eddies present in the wake also contribute to heat and mass transport across the flame surface. The higher burning velocity increases the flow velocity in the unburned mixture, which creates more turbulence, leading to a further increase in burning velocity, and so on. With periodic obstacle spacing, the positive feedback mechanism between the flow dynamics and the burning velocity leads to rapid flame acceleration.

Drawing on the relevant discussion of turbulence and burning velocity given in Chapter 14, the preceding explanation by Moen et al. [13] can be applied to dust deflagrations. It is also consistent with the case of post-ignition turbulence generation caused by combustion product expansion and wall effects, as described in Section 16.1. Obstacle-induced turbulence is not, however, restricted to theoretical considerations and research laboratories. The matter can have quite practical implications.

Both Eckhoff [14] and Abbasi and Abbasi [15] have commented on the possibility of the buckets in a bucket elevator (see Figure 17.3) acting as turbulence-generating obstacles should a dust explosion propagate through one of the transfer legs. Similarly, Figure 17.4 shows a post-explosion view of a boom truck (essentially a tractor for moving heavy items underground) in the Westray coal mine. (See Chapters 4 and 21.) This vehicle was positioned close to the likely origin of the initial methane explosion [16] and hence shows evidence of burning with limited structural damage. Wire support mesh for the ventilation air ducting with burned remnants of the covering can be seen on the right side of Figure 17.4. No one, of course, can say with certainty whether the boom truck acted as a turbulence generator leading to flame acceleration in the mine corridor. Nevertheless, the implications from our understanding of basic physics and chemistry should be clear.

FIGURE 17.3   The uptake leg of a bucket elevator with an open panel showing the individual buckets.

FIGURE 17.4   A boom truck in the Westray coal mine [16].

17.2 TEMPORARY ENCLOSURES

One of the first incident investigations conducted by the U.S. Chemical Safety Board involved nitrogen asphyxiation at a Union Carbide plant in Hahnville, LA, causing the death of one worker and serious injury to another. All incident details that follow are drawn from the CSB investigation report [17].

The men were performing an inspection for organic residue such as grease or oil on the flange at the open end of the 48-inch wide horizontal pipe shown in Figure 17.5. Unknown to these men, nitrogen was being purged through the pipe. The particular inspection technique being used relied on the use of "black light" that would cause the undesired residue to shine and therefore be more readily identified for further cleaning. A sheet of black plastic was secured over the end of the pipe to provide shade from the midday sun, thus facilitating the observation of residue. The workers were positioned inside the black plastic sheet and just outside the pipe opening. They were overcome by the nitrogen venting from the pipe. (See Chapters 10 and 11.)

Although I no longer recall the specific reference, I remember reading some years ago of a similar incident in which welders were working outdoors near the open end of a large pipe. Because it was raining, they erected plastic sheeting around the pipe opening to provide protection from the inclement weather. Welding operations proceeded with the inevitable accumulation of welding gases and fumes. Again, injuries to personnel occurred.

What these two incidents have in common is the creation of a confined space by means of a temporary enclosure. Increased risk from specific gas-phase hazards then ensued.

Could the same use of temporary structures increase the level of confinement in a dust-handling facility to the point where the likelihood and/or severity of a dust explosion is similarly increased? The answer would seem to depend on, among other factors, the strength of the temporary enclosure and its proximity to dust

FIGURE 17.5 Pipe opening (48-inch) with black plastic sheet lying nearby [17].

accumulations. I am not personally aware of this situation arising in industry—but, of course, that does not mean it hasn't happened or that it might not in the future.

My intention is certainly not to raise a false alarm with these thoughts. It's just that a key causal factor in confined space incidents such as the nitrogen and welding examples given here has been a well-intentioned motivation to achieve greater efficiency in a work process or to shelter oneself from the weather. This is a sobering reminder of the importance of hazard identification in even the seemingly benign aspects of the process industries.

17.3 REALITY

The development of destructive overpressures does not occur only with confinement arising from the structural boundaries provided by equipment and building walls. We have therefore seen in this chapter the need to think of confinement in terms of the actual role it plays in the explosion pentagon.

Post-ignition or explosion-generated turbulence is made possible by workplace congestion and obstacles that increase the ratio of blocked to open area. Dust combustion more rapid than occurs in the absence of such features leads to flame acceleration processes and potentially high explosion pressures. Although dust detonations are rare in industry, their occurrence is not precluded should conditions of dust reactivity and flow geometry enable the transition from deflagration flame speeds.

When one thinks of potential confinement scenarios, the possibility of a temporary structure creating a confined or semi-confined space must also be considered. Industrial case studies have demonstrated the risk from asphyxiation or inhalation of noxious fumes from activities occurring in such enclosures. The issue of whether a dust explosion could occur in a temporarily and unintentionally created confined area should not be automatically dismissed.

17.4 WHAT DO *YOU* THINK?

Taking a global view of the material in Chapters 16 and 17, we could come up with the following list of factors that can create a level of confinement conducive to generation of dust explosion overpressures:

- Completely enclosed process vessels
- Completely enclosed buildings and other work spaces
- Long corridors or other semi-enclosed spaces in the work environment
- Undersized explosion relief vents
- Vent ducts
- Obstructions placed on or near vents
- Congestion in an otherwise more-or-less open environment
- Turbulence-generating obstacles in the workplace
- Temporary enclosures leading to increased levels of confinement and perhaps even the creation of a confined space.

Are there other general confinement scenarios you would add to the preceding list? Using the list as a checklist of sorts, can you identify specific cases of confinement where combustible dusts are handled in your workplace? Recognizing that confinement is a key factor in the transition from a flash fire to an explosion, do you have an appropriate mix of prevention and mitigation measures in place for each specific case identified?

Finally—and reflecting back on the need for management of change as detailed in Section 11.3—have you made any changes in your facility that have altered the degree of confinement with respect to your handling of combustible dusts? Have you considered the effects of these changes on the dust explosion control measures in place?

On this last point, a relevant case study mentioned in Section 11.5 is the 2008 explosion at the Imperial Sugar Company in Port Wentworth, GA. (See also Chapter 3.) Note the following excerpt from the U.S. Chemical Safety Board investigation report [18] (p. 29) on modifications made to the steel belt conveyor at the facility:

The CSB investigators learned through worker interviews that conveying granulated sugar on the steel belt conveyor in the tunnel under silos 1 and 2 generated some sugar dust. When sugar lumps lodged in a silo outlet pipe and blocked the movement of the sugar on the belt, as operators reported sometimes occurred, sugar spilling off the belt would release airborne dust. Before the company enclosed the steel belt conveyor, the dust was released into the large volume of the tunnel. Airflow through the tunnel would also keep the airborne dust concentration low; thus, airborne sugar dust likely never accumulated to an ignitable concentration.

In 2007, Imperial Sugar enclosed the granulated sugar belt conveyors in the penthouse and the tunnel under the silos to address sugar contamination concerns. However, the company did not evaluate the hazards associated with generating and accumulating combustible dust inside the new enclosure. They did not install a dust removal system to ensure that sugar dust did not reach the MEC [minimum explosible concentration] inside the enclosure. Furthermore, the enclosures were not equipped with deflagration vents to direct overpressure safely out of the building if airborne sugar dust ignited.

REFERENCES

[1] Crowl DA. Understanding explosions. New York, NY: Center for Chemical Process Safety, American Institute of Chemical Engineers; 2003.

[2] Eckhoff RK. Dust explosions in the process industries, 3rd ed. Boston, MA: Gulf Professional Publishing/Elsevier; 2003.

[3] Hertzberg M, Cashdollar KL. Introduction to dust explosions. In: Cashdollar KL, Hertzberg M, editors. Industrial dust explosions. ASTM Special Technical Publication 958. Philadelphia, PA: American Society for Testing and Materials; 1987. pp. 5–32.

[4] Abbasi T, Pasman HJ, Abbasi SA. A scheme for classification of explosions in the chemical process industry. Journal of Hazardous Materials 2010;174:270–80.

[5] James H. Detonations. TD5/039 Bootle, UK: Health and Safety Executive; 2001 (Available at http://www.hse.gov.uk/foi/internalops/td_din/fire-exp/td5_039/td5_039.htm; last accessed November 20, 2012).

[6] Exponent. Explosions, deflagrations & detonations. Exponent Engineering and Scientific Consulting. 2012 (Available at: http://www.exponent.com/explosions/; last accessed November 20, 2012).

[7] Sichel M, Kauffman CW, Li YC. Transition from deflagration to detonation in layered dust explosions. Process Safety Progress 1995;14:257–65.

[8] Amyotte PR, Chippett S, Pegg MJ. Effects of turbulence on dust explosions. Progress in Energy and Combustion Science 1988;14:293–310.

[9] Lunn GA, Holbrow P, Andrews S, Gummer J. Dust explosions in totally enclosed interconnected vessel systems. Journal of Loss Prevention in the Process Industries 1996;9:45–58.

[10] Holbrow P, Andrews S, Lunn GA. Dust explosions in interconnected vented vessels. Journal of Loss Prevention in the Process Industries 1996;9:91–103.

[11] Kosinski P, Hoffmann AC. An investigation of the consequences of primary dust explosions in interconnected vessels. Journal of Hazardous Materials 2006;A137:752–61.

[12] Zhou Y, Bi M, Qi F. Experimental research into effects of obstacle on methane-coal dust hybrid explosion. Journal of Loss Prevention in the Process Industries 2012;25:127–30.

[13] Moen IO, Donato M, Knystautas R, Lee JH. Flame acceleration due to turbulence produced by obstacles. Combustion and Flame 1980;39:21–32.

[14] Eckhoff RK. Understanding dust explosions. The role of powder science and technology. Journal of Loss Prevention in the Process Industries 2009;22:105–16.

[15] Abbasi T, Abbasi SA. Dust explosions—cases, causes, consequences, and control. Journal of Hazardous Materials 2007;140:7–44.

[16] Richard KP, Justice. The Westray story—a predictable path to disaster. Report of the Westray Mine Public Inquiry. Halifax, NS, Canada: Province of Nova Scotia; 1997.

[17] CSB. Summary report—nitrogen asphyxiation—Union Carbide Corporation. Report No. 98-05-I-LA Washington, DC: U.S. Chemical Safety and Hazard Investigation Board; 1999.

[18] CSB. Investigation report—sugar dust explosion and fire—Imperial Sugar Company. Report No. 2008-05-I-GA Washington, DC: U.S. Chemical Safety and Hazard Investigation Board; 2009.

Chapter 18

Myth No. 17 (Pentagon): The Vocabulary of Dust Explosions Is Difficult to Understand

The previous 16 chapters each dealt primarily with one specific element of the explosion pentagon. (Chapter 7 covered aspects of both *fuel* and *ignition source*.) In the next four chapters, we will address some myths and realities that are more general in nature and which therefore can be considered to involve all five elements of the explosion pentagon.

As the title states, the subject of the current chapter is the notion that the vocabulary of dust explosions is difficult to understand. This is a feeling that has been expressed to me numerous times over the years by people with various backgrounds and varying degrees of scientific/engineering expertise—often with the addition of *too* in front of *difficult*.

I empathize with the frustration inherent in such expressions. The vocabulary of dust explosions is indeed different in some respects from that typically found in the occupational health and safety (OHS) field. Yet as we will explore soon, it is also similar to the concepts and terms used when dealing with gas explosions. Nevertheless, if one is not an "expert" in the dust explosion field, the vocabulary of dust explosions might be viewed as prohibitively difficult to understand. Or as an individual once commented to me when I offered to provide a training session to his workforce: "It's too technical for our people."

Although I have empathy for those who feel dust explosion terminology is somehow different or difficult to comprehend, I have no sympathy for managers who ignore their responsibility to educate their workers on the hazards and risks of dust explosions. I have had the experience of working for and providing testimony to the Westray Mine Public Inquiry [1] at a relatively early stage in my professional career. That experience has clearly shaped my thoughts on the matter.

18.1 DUST EXPLOSION TERMINOLOGY

As articulated in Section 15.5, dust explosions are process safety incidents—with *process safety* being defined as the prevention and mitigation of process-related injuries and damage arising from process incidents (fire, explosion, and toxic release). To understand what is meant by *process,* we can turn to the U.S. Occupational Safety and Health Administration (OSHA), which has defined a process as any activity involving a highly hazardous chemical including any use, storage, manufacturing, handling, or the on-site movement of such chemicals, or combination of these activities [2].

A word of caution though—dust explosions occur in many industries, not just the traditional process (or chemical process) industries. Yes, they occur in petrochemical plants where the chemical (e.g., polyethylene) is the desired product; but they also occur in powdered metallurgy manufacturing facilities where the dust (e.g., iron) may not be viewed as a "hazardous chemical," or in a coal mine where the same thought might apply to nuisance coal dust. (Recall the broad range of industrial products and processes detailed in Chapter 3.)

I am not attempting to redefine the process industries with these comments. What I am doing is making the point that regardless of the industry, it is the application of *process* safety principles that will be most effective in reducing the risk of a dust explosion. This means, among other items, the application of the hierarchy of controls for *process* hazards (Chapter 15), the adoption of a *process* safety management system (Chapter 21), and the development of a *process* safety culture (Chapter 21). The dust explosion problem is not going to be solved by an emphasis on occupational safety alone.

Occupational or "traditional" safety largely aims to control individual or personal exposures—often referred to as slips, trips, and falls. Protective devices such as machine guards, the wearing of personal protective equipment (PPE), and safe work practices such as confined space entry and lockout/tagout are critical features of occupational safety. OHS principles are absolutely essential to safer industrial practice and have an obvious overlap with enhancing process safety; but again, their focus of application is largely on the individual worker.

This fundamental difference between process safety and occupational safety becomes especially evident when examining the familiar *incident pyramid* shown in Figure 18.1. (The discussion here follows that given by Rigas and Amyotte [2].) Various studies over the years, covering a broad range of industries and typically from an OHS perspective, have resulted in different category totals for the pyramid levels but generally the same ratios between levels. (See, for example, Bird and Germain [3].) As explained by Creedy [4]:

The idea of the pyramid is that really serious incidents such as fatalities occur so rarely in most organizations these days that it's not practical to use them as a measure for monitoring and improving an organization's safety effectiveness—there simply aren't enough such incidents to know whether things are getting better or worse. However, it is practical to track the larger number of less serious incidents and use that as a performance measure.

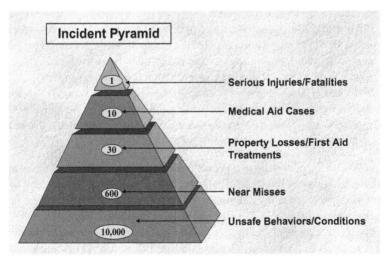

FIGURE 18.1   The incident pyramid [4].

Hence, many safety programs—particularly those dealing with OHS and avoidance of lost-time injuries (LTIs)—rightfully emphasize the prevention, control, and investigation of near-misses and unsafe (or at-risk) acts and conditions. Creedy [4] does, however, provide a cautionary note:

The pyramid is certainly a useful concept. Nevertheless, it does have a serious problem that is often unrecognized, even by workplace health and safety professionals. The problem is that the conditions that can lead to really serious incidents—those which could kill or seriously injure a large number of people—may not be identified by focusing on the bottom end of the pyramid, which could thus become a distraction rather than a help.

Thus, if a safety program is designed to facilitate effective process safety efforts, investigation of near-misses must involve process-related near-misses such as the transient formation of a dust cloud in an occupied space, overpressurization incidents inside process vessels, electrostatic sparking with no ignition, etc. Investigations of OHS near-misses involving working at height, for example, are ineffective and may be counterproductive if the desired focus is process safety. (These measures are, of course, entirely appropriate if the focus is on OHS.)

It may be helpful at this point to provide some historical perspective. In his recounting of safety philosophy development in the process industries, Creedy [4] describes four phases of development:

- *Late 19th and early 20th century*—Here, the objective was primarily the protection of capital assets, providing the origins of basic safety thinking in the explosives industry.
- *Second World War through the 1950s and 1960s*—During this period, the objectives were greater efficiency and the creation of a better society. The concepts of loss prevention and investing in people were introduced. Safety measures were largely rule-based.
- *1970s and 1980s*—The objectives here were the same as for the previous phase. However, recognition of consequence seriousness and causation mechanisms led to a focus on the process rather than the individual worker, and hence the development of a management approach to process safety.
- *1990s and beyond*—Here we see a realization of the significance of sociocultural factors in human thought processes and behaviors at the individual and organizational levels. This has led to increased understanding of the importance of such concepts as human factors and safety culture.

At present, we appear to be in a period that merges the latter two of the preceding phases—process safety management coupled with a recognition that without a strong process safety culture, even the best management system on paper can become dysfunctional. We will return to this point in Chapter 21.

The distinction made in this section between process safety and occupational safety is far from a simple academic exercise. Effectively addressing the issue of industrial dust explosions means incorporating the language and principles of process safety into our collective approach. It means gaining familiarity with the terminology given in Table 5.1 for often-encountered dust explosibility parameters—and knowing which of these parameters relate to explosion likelihood and which to explosion consequence severity, as explained in Section 5.1. The acronyms are fairly standard and, to a large extent, are used consistently throughout the world.

A process safety approach to risk reduction also means recognizing that while personal protective equipment such as fire-retardant clothing may protect a worker from some of the consequences of a dust explosion, it does nothing to prevent a dust explosion. The measures given in Table 15.2—inherently safer design, inerting, venting, suppression, explosion isolation, and the like—are needed to prevent dust explosions and mitigate their consequences.

18.2 GAS EXPLOSION ANALOGIES

As speculated in Chapter 2, the word *gas* may impart a more general sense of hazard awareness than *dust*. I also remarked that there appears to be a stronger linkage between the words *gas* and *explosible* than between the solid counterpart *dust* and *explosible*. To assist with understanding the terminology of dust explosions, one therefore may find it beneficial to look to the field of gas explosions. Consistent with the material in Chapter 5, our purpose is to learn from the similar parameter names, not to make comparisons as to which fuel type is "worse."

All but one of the parameters in Table 5.1 have gas-phase counterparts; the layer ignition temperature (LIT) is unique to dust explosions. For example, the minimum explosible concentration (MEC) can be thought of as the solid-phase equivalent to the gas-phase lower flammable limit (LFL). As I commented earlier in Chapter 2, my preference is to use *explosible* for the E in *MEC*, leaving *explosive* to be used with materials intended for overpressure generation (i.e., explosives such as dynamite). Others may adopt a different convention, but there should be little confusion. It's a bit like using *lean* rather than *lower* for the first L in *LFL*.

Further, the gas-phase equivalents of the parameters in Table 5.1 are also determined via standardized testing using standardized apparatus. A significant difference, though, between gas and dust explosibility testing is the lack of a turbulent air blast to create a fuel/air suspension with the former. This is not to say that gas explosion testing is without complication; witness the fact that gas-phase flammability limit ranges are wider for upward flame propagation than for downward flame travel and thus represent a more conservative hazard indicator.

Several references have been made in this book to K_{St}, the size-normalized maximum rate of pressure rise for a dust. While further details on the determination of this parameter are given in the next chapter, we know already that key test features include the dispersing air pressure, ignition delay time, and ignition energy. Again, measurement of K_G (the gas-phase counterpart to K_{St}) would not involve such issues. K_G determination does, however, require consideration of the test vessel size. For very small vessels, the flame may not be fully developed before quenching occurs at the vessel walls. For large vessels, hydrodynamic instabilities caused by the interaction of reflected pressure waves with the flamefront can act as an alternate flame acceleration mechanism to turbulence and essentially invalidate the concept of K_G [5].

The prevention and mitigation of explosions of flammable gases therefore rely on many of the same parameters as when dealing with explosions of combustible dusts—all determined according to standardized protocols having their intrinsic concerns. Gas explosions are, I believe, reasonably well understood in terms of the vocabulary used to describe their origin and effects. Is there any justifiable reason for the field of dust explosions to fall behind in this regard?

18.3 RIGHT TO KNOW

On April 27, 2012 (Workers Memorial Day in the United States), Dr. Rafael Moure-Eraso, Chairperson of the U.S. Chemical Safety Board, commented that "a safe workplace is a basic human right [6]." This seems to me to be a simple and eloquent call to action.

One way to ensure our workplaces are indeed made safer is to educate all members of the workforce on the hazards present in their daily tasks. As expressed in a recent European Union publication [7] (p. 9):

Employers must give workers information about the risks in their workplace and how they are protected, and also instruct and train them on how to deal with the risks.

Safety regulations in my home province of Nova Scotia, Canada, make explicit reference to workers' *right to know* about hazards in the workplace and how to protect themselves from such hazards. This right is expressed in the form of employers' precautions and duties, two of which are to [8]

provide such information, instruction, training, supervision and facilities as are necessary to the health or safety of the employees,

and

ensure that the employees, and particularly the supervisors and foremen, are made familiar with any health or safety hazards that may be met by them at the workplace.

It is widely accepted in Nova Scotia that these regulations are a direct result of the Westray coal mine explosion. (See Chapters 4 and 21.)

18.4 REALITY

There is a distinct need to employ process safety concepts and approaches in attempting to reduce the risks of dust explosions. These events tend to fall in the category of lower frequency/higher consequence incidents for which traditional personal safety measures alone are largely inadequate.

The vocabulary of dust explosions is similar to that used for gas-phase explosions. Whether the terminology is viewed as being difficult to comprehend is not a valid reason to dismiss the obligation an employer holds to communicate hazards to employees in clear, understandable language.

It is suggested that a differentiated approach to training in the fundamentals of dust explosion hazards would be beneficial. All workers involved in the handling of combustible powders should know the basic terminology as expressed by the parameters in Table 5.1. Not everyone in the workplace, however, needs to know the details of K_{St} determination (as discussed in the next chapter), or how to use such data for sizing explosion relief vents. It may be sufficient in some cases to know that K_{St} is a measure of explosion severity or violence, and that a K_{St} of 200 bar·m/s indicates a faster rate of pressure rise than does a K_{St} of 100 bar·m/s. Similarly, concepts such as MIE and MIT can be related to actual spark energies and equipment surface temperatures, respectively, to provide process relevance to the explosibility data for a given material.

18.5   WHAT DO *YOU* THINK?

"It's too technical for our people." I have often wondered what motivated this comment I mentioned in the introductory portion of the current chapter.

What motivates you to provide combustible dust hazard awareness training to your employees? If you currently work in a facility or industry that does not handle combustible dusts, what motivates you to provide training commensurate with the hazards you do face?

Quoting from Chapter 1:

The writing of this book has been motivated in equal measure by a desire to aid in the protection of people, business assets, operational production, and the natural environment, and a need to address important communication issues with respect to understanding dust explosions.

The first part of the preceding passage is an expression of the integrated approach to loss prevention advocated by Wilson and McCutcheon [9]. This concept is often stated as a desire to avoid loss to people, property, production, and the environment. Is there motivation here to prevent and mitigate dust explosions?

Taking a slightly different approach, there are many arguments for attempting to ensure safer workplaces: (i) moral, (ii) ethical, (iii), legal, and (iv) financial. I make a distinction between *moral* and *ethical* reasons because for some, the motivation goes beyond the moral persuasion of acting simply because it is the *right thing to do*. For some, a code of ethics will also be applicable. In my own case as a professional engineer (P.Eng.) in the province of Nova Scotia, I am bound by the code of ethics adopted by the Association of Professional Engineers of Nova Scotia [10]. This code has as its first basic tenet the requirement to "hold paramount the safety, health and welfare of the public and the protection of the environment and [to] promote health and safety within the workplace."

Do you have a similar ethical mandate to ensure combustible dust hazards are well known within your workplace? Are there specific legal requirements in your geographical region? Have you considered the financial implications of a dust explosion in your plant?

Finally, there is the matter of *reputation*. Some may consider this to be captured under the arguments already presented—perhaps as an aspect of the financial category. But let's break it out separately given the high-profile incident that occurred recently in the Gulf of Mexico offshore oil industry. Can your company withstand the reputational damage that might accompany a dust explosion in one of your facilities?

It is hopefully obvious that I would rank the *moral and ethical motivation for avoiding loss to people* above all else in this discussion of why dust explosion hazards must be clearly communicated.

REFERENCES

[1] Richard KP, Justice. The Westray story—a predictable path to disaster. Report of the Westray Mine Public Inquiry Halifax, NS, Canada: Province of Nova Scotia; 1997.

[2] Rigas F, Amyotte P. Hydrogen safety. Boca Raton, FL: CRC Press, Taylor & Francis Group; 2012.

[3] Bird FE, Germain GL. Practical loss control leadership. Loganville, GA: DNV; 1996.

[4] Creedy G. Process safety management. Ottawa, ON: PowerPoint presentation prepared for Process Safety Management Division, Chemical Institute of Canada; 2004.

[5] Amyotte PR, Chippett S, Pegg MJ. Effects of turbulence on dust explosions. Progress in Energy and Combustion Science 1988;14:293–310.

[6] Moure-Eraso R. Statement from U.S. Chemical Safety Board Chairperson Rafael Moure-Eraso on Workers Memorial Day. Washington, DC: U.S. Chemical Safety and Hazard Investigation Board; 2012.

[7] European Agency for Safety and Health at Work. Worker participation in occupational safety and health. A practical guide. Luxembourg: Publications Office of the European Union; 2011. (Available at: http://osha.europa.eu/en/publications/reports/workers-participation-in-OSH_guide; last accessed November 25, 2012.)

[8] Occupational Health and Safety Act. Chapter 7 of the Acts of 1996. An act respecting occupational health and safety. Halifax, NS: Province of Nova Scotia; 1996. (Available at:http://nslegislature.ca/legc/statutes/occph_s.htm; last accessed November 25, 2012.)

[9] Wilson L, McCutcheon D. Industrial safety and risk management. Edmonton, AB, Canada: University of Alberta Press; 2003.

[10] Engineers Nova Scotia. Code of ethics. Halifax, NS: Association of Professional Engineers of Nova Scotia; 2012. (Available at:https://www.engineersnovascotia.ca/uploads/Code_of_Ethics_2012.pdf; last accessed November 25, 2012.)

Myth No. 18 (Pentagon): Dust Explosion Parameters Are Fundamental Material Properties

The result of testing for the explosibility parameters listed in Table 5.1 is, of course, *data*—i.e., *numbers*. It is tempting, perhaps understandably, to view such numbers as physical properties that are constant for a given material. In this chapter we will examine why this is not the case—why we cannot consider parameters such as P_{max}, $(dP/dt)_{max}$, K_{St}, MEC, MIE, MIT, LIT, and LOC to be fundamental material properties of a dust. The chapter is a summary of sorts because we have already seen most of the reasons through previous discussion in this book.

If you were asked to state the boiling point of water, you would no doubt respond with *100 °C* (or perhaps *212 °F* depending on your geographic location). You might think to ask about the ambient pressure, but more likely you would assume the questioner meant 1 atmosphere pressure. And if you were asked for the specific heat capacity of liquid water, you might respond with *4.184 J/g·K* (or a single value using different units), without worrying about any temperature dependence. But if asked for the density of air, you would be unable to answer unless told the pressure and temperature at which to determine the value. Even then, you would need a pressure-volume-temperature (PVT) relationship appropriate for the conditions.

It's not a perfect analogy, but in a similar fashion, it is not possible (in the absence of other information) to answer a question like: *What is the maximum rate of pressure rise of polyethylene?* Again, in accordance with previous chapters, such a parameter is strongly dependent on both material characteristics (e.g., moisture content, particle size, and particle shape) and experimental conditions (e.g., vessel volume, turbulence level, and applied ignition energy).

And, of course, we know well that determination of dust cloud explosibility requires the generation of a dust/oxidant suspension, usually by means of a compressed air blast. The concept of an initially *quiescent* dust cloud therefore has no physical meaning within the confines of the earth's gravitational field. But what about in space?

19.1 A QUIESCENT DUST CLOUD—THE (NEARLY) IMPOSSIBLE DREAM

The first sentence of Agarwal's practical and informative paper [1] contains this phrase (p. 26): "preventing dust explosions is not rocket science." This may be true figuratively speaking in that we do seem to have available many of the practical tools needed for dust explosion prevention and mitigation. But in a literal sense, rocket science may be exactly what is needed to further our fundamental understanding of the nature of dust flame propagation processes—thereby providing information that is essential for a less empirical and more analytical/computational approach to dust explosion risk reduction.

As explained in Section 14.4, the laminar burning velocity (S_L) is the flame propagation velocity relative to the unburned fuel/air mixture under laminar conditions. This parameter is an important and fundamental characteristic of a given fuel. Laminar burning velocities have been measured for many flammable gases using either constant-volume (spherical vessel) or constant-pressure (burner) techniques [2]. There have been several laminar or quasi-laminar flame propagation studies undertaken for combustible dusts using burners and drop towers [3]. Laminar burning velocities have also been estimated by calculation from pressure/time traces determined in constant-volume dust explosion tests [4].

Bind et al. [5] note, however, that recent modeling efforts using a computational fluid dynamics (CFD) approach and data from laboratory-scale testing have demonstrated the need for better understanding of the basic physics and chemistry of dust explosions. Industrial scenarios involve turbulent dust clouds, and predictive methods therefore require a measure of the turbulent burning velocity—which is typically expressed as a function of the laminar burning velocity.

The laminar burning velocity thus represents a key building block for models of dust flame propagation. Yet we are left with the familiar issue that, by definition, this *laminar* parameter cannot be directly measured in a *turbulent* dust cloud. Hence, we can gain an appreciation for the value of combustible dust studies under conditions of reduced gravity, such as those conducted by Goroshin et al. [3] on a parabolic flight aircraft. Figure 19.1 shows propagating flames from their work with iron dust samples of different particle size and shape. It is not often one reads passages such as the following [3] (p. 659) in the dust explosion literature (with underlining added for emphasis):

The experimental sequence consisted of a dust dispersion period terminated after 8–10 s by the cut-off of dispersion flow, a 0.5–2 s ignition delay, and a 3–10 s flame propagation event through <u>the quiescent dust suspension</u>.

As with the brief discussion of detonations in Chapter 17, it is beyond the scope of this book to delve much further into the science of dust combustion. For our purposes here, it is sufficient to know that while there are indeed fundamental flammability/explosibility parameters in the field of dust explosions, they are (i) challenging to determine, and (ii) not the parameters given in Table 5.1 (P_{max}, K_{St}, etc.). We are also now in a position to recognize that a call for further study of dust flames in low levels of gravity on orbital platforms [3] is not simply science fiction. And finally, we can better understand both the role of parameters such as P_{max} and K_{St}, and the quest for more fundamental measures of explosibility, as expressed by Zalosh [6] in his comprehensive review of explosion venting (p. 5 and p. 37, respectively):

A premise in most of the contemporary data correlations is that the dust explosibility can be represented in terms of the K_{St} value and the P_{max} value...of the material, for a given particle size distribution.

and

The phenomenological and CFD models currently being developed to simulate dust explosions use material properties that are much more fundamental than the measured value of K_{St} for a particular dust. Therefore, they should eventually be quite useful for analyzing the effects of non-standard dust dispersion and turbulence conditions in vented dust explosions.

(See also Section 14.4.)

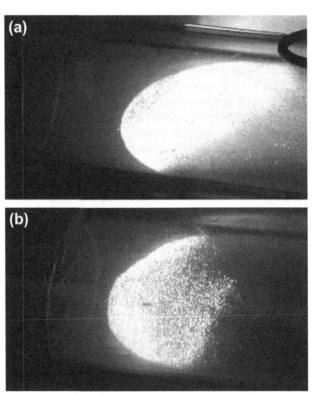

FIGURE 19.1 Propagating flames in reduced-gravity suspensions of iron dust: (a) spherically shaped particles having an arithmetic mean diameter of 7.0 μm, and (b) irregularly shaped particles having an arithmetic mean diameter of 26.8 μm [3].

19.2 THE MYSTICAL K_{St} PARAMETER

Over the past 20 years of providing explosibility data to industry, I have found that K_{St} is the one parameter that has consistently proven the most difficult for me to explain and for my clients to understand. Hence, I have labeled K_{St} as *mystical* in the title of this section to indicate that some people may see it as somewhat cryptic—or just plain baffling. I would hasten to add that I have always felt the onus for clear communication in these matters lies with me; the responsibility for any shortcoming similarly lies with me. The current section and the suggested exercise in Section 19.5 are an attempt to demystify this parameter that is so extremely important for protection from the consequences of a dust explosion.

Let us first return to the definition of K_{St} given in Section 6.2. The maximum rate of pressure rise itself, $(dP/dt)_{max}$, is, of course, dependent on the volume of the explosion chamber, and is therefore of limited use on its own. For scaling to larger volumes, maximum rates of pressure rise are normalized by multiplying by the cube-root of the explosion chamber volume, V:

$$K_{St} = (dP/dt)_{max} \cdot V^{1/3} \tag{19.1}$$

Equation 19.1 is sometimes referred to as the cubic or cube-root *law* and K_{St} as the *dust constant*. (The subscript *St* derives from the German word for dust—*staub*.) It is preferable, however, to refer to Equation 19.1 as the cubic *relationship* and K_{St} as the *size- or volume-normalized (or standardized) maximum rate of pressure rise* (Table 5.1), or simply as K_{St}. There is nothing fundamental (in the sense of an inviolable law) or constant about Equation 19.1, or—as we have previously seen—the K_{St} parameter.

This observation is reinforced by an adaptation of the discussion given by Eckhoff [7], in which the need for appropriate determination of K_{St} values is emphasized. An analysis of basic considerations reveals that the cubic relationship is valid only (i) for geometrically similar vessels giving geometrically similar flame surfaces, (ii) if the flame thickness is negligible compared to the vessel radius, and (iii) if the burning velocity as a function of pressure and temperature is identical in all volumes. In view of these issues and previous discussion in Chapters 14 and 18, it is clear that a value of $(dP/dt)_{max} \cdot V^{1/3}$ from a dust explosion in any arbitrary vessel is a correspondingly arbitrary measure of dust explosion violence. Both the turbulence level and the vessel shape are arbitrary, and the flame thickness is most probably considerable in relation to the vessel radius.

Table 19.1 shows a selection of $(dP/dt)_{max} \cdot V^{1/3}$ values for maize starch dust clouds in air, determined in various apparatus. (Readers can consult Eckhoff [7] for the listing of investigators and original references.) The values range from 3 to 6 bar·m/s to over 200 bar·m/s, corresponding to approximately an order of magnitude difference. Some of the discrepancies may arise from differences in moisture content and effective particle size of the starch, and to different data interpretation (peak or mean values). However, differences in turbulence levels of the dust clouds and significant flame thicknesses probably play the main roles [7].

Therefore, when one is using K_{St} values for sizing of explosion vents and for design of explosion isolation and explosion suppression systems according to current standards, it is absolutely essential to use data obtained from authorized standard test methods for determining K_{St}. This point has been previously made on several occasions in the current book; Figure 19.2 illustrates the data reproducibility possible with standardized K_{St} testing. Here, we see the results of K_{St} determination for the same dust as that conducted in laboratories in Asia, Europe, and North America. The mean K_{St} was measured as 243 bar·m/s, with the vast majority of participants reporting values falling within ± 10% of the mean value, as shown in Figure 19.2.

TABLE 19.1 Maximum Rates of Pressure Rise Measured for Clouds of Maize Starch Dust in Air in Different Closed Vessels and Arranged According to Vessel Volume

$(dP/dt)_{max}$ [bar/s]	Volume (V) of apparatus [m³]	$(dP/dt)_{max} \cdot V^{1/3}$ [bar·m/s]
680	0.0012	73
612	0.0012	66
220	0.0012	23
413	0.009	86
320	0.020	87
365	0.020	100
10–20	0.026	3–6
60–80	0.026	20–25
272	0.028	83
50	0.33	34
72	0.95	71
20	0.95	20
136	3.12	200
110	6.7	209
55	13.4	131

(adapted from Eckhoff [7])

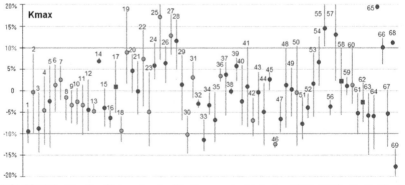

FIGURE 19.2   Results of international calibration-round-robin testing (CaRo 11) for K_{St} (or K_{max}) [8]. The circles represent data from 66 standardized 20-L chambers (using chemical ignitors from either of two manufacturers); the squares represent data from three standardized 1-m³ chambers.

The Kuhner website [8] can be consulted for further details on the calibration-round-robin testing leading to the results shown in Figure 19.2. Of interest would be mention of the calibration step of rectifying any issues for laboratories falling outside the tolerance range of ±10% of the mean K_{St}. The calibration exercise also extended to reporting of standardized P_{max} and MIE values; these results are available on the Kuhner website [8].

To close this section, we return to another quote from the venting report by Zalosh [6] (p. 5):

Although K_{St} is not a fundamental property of a combustible dust, it has become by far the most widely used experimental parameter to characterize material explosibility under prescribed, reproducible test conditions.

Figure 19.2 helps to explain why this is the case.

19.3 STANDARDIZED DUST EXPLOSIBILITY TESTING (REVISITED)

Barton [9] comments that in addition to turbulence, a secondary factor influencing the rate of pressure rise during a dust explosion is the strength of the ignition source. Previous sections in the current book have highlighted the connection between ignition energy (Chapter 8) and turbulence level (Chapter 14). Generally speaking, laboratory-scale MIE testing involves low turbulence and low ignition energies, whereas P_{max}/K_{St} testing involves considerably higher values of both factors. Laboratory-scale testing at the limits of explosibility for parameters such as MEC and inerting level is typically done using the same high turbulence intensity as for P_{max}/K_{St} but with a lower ignition energy.

Figure 19.3 quantitatively demonstrates the influence of ignition energy when testing for inerting level with a standard 20-L chamber [10]. (See Chapter 7.) Mine-size coal dust was inerted with rock dust (dolomite in this case) using ignition energies ranging from 0.25 to 20 kJ and a fixed coal dust concentration of 500 g/m³ [11]. The explosion overpressure decreased for all ignition energies tested as rock dust concentration was increased; when enough rock dust was added, the mixture was completely inerted and explosions were no longer possible [as evidenced by the overpressure dropping below the explosion criterion of 1 bar(g)]. A wide range of inerting levels exists in Figure 19.3—from just under 50 mass % dolomite at an ignition energy of 0.25 kJ to just over 80 mass % dolomite at an ignition energy of 20 kJ. Inerting level is clearly not an invariant material property, and the selection of an ignition energy for its determination in the laboratory is entirely dependent on comparison with full-scale mine test results.

The explanation for the behavior shown in Figure 19.3 rests with the phenomenon known as *overdriving*, which was introduced in Section 16.4. The use of energetic ignition sources in relatively small vessels has been shown to cause preconditioning of the combustion volume in terms of increases in fluid

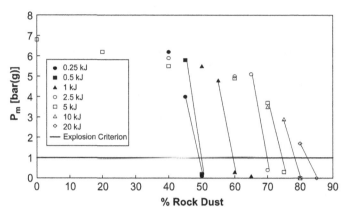

FIGURE 19.3   Effect of ignition energy on the inerting level of mine-size coal dust inerted with rock dust in a Siwek 20-L chamber [10].

pressure and temperature as well as alteration of particle temperature and concentration prior to actual flame propagation [12]. Overdriving has long been a recognized factor in dust explosion testing and provides yet further evidence of the fact that explosibility parameters are non-fundamental artifacts of the test conditions used in their determination.

In the case of P_{max} and K_{St}, the usual reference scale for 20-L test results is a volume of 1 m^3. Values determined using a 20-L chamber generally compare favorably with those measured in a 1-m^3 chamber when both vessels are operated under standardized test conditions. A notable exception is what have been termed *marginally explosible dusts*; these are materials with low K_{St} values of the order of 50 bar·m/s as determined in a 20-L volume. The concern here is the occurrence of false positives whereby such dusts would not explode in the larger 1-m^3 volume, their apparent explosibility at the scale of 20 L being due solely to overdriving. The implicit assumption, which is in fact valid, is that energetic ignition sources do not have the same preconditioning effects on large test volumes.

A recent paper by Bucher et al. [13] is an excellent entry-point for readers who may wish to pursue this topic in more detail. These authors give a good summary of overdriving concerns and also relate the development history of the standard 20-L chamber named for its designer, Richard Siwek. Further, Bucher et al. [13] provide empirical evidence that while not all low K_{St} (20-L) dusts explode in a 1-m^3 chamber, some—especially metals—can actually yield higher values of explosion pressure and rate of pressure rise in the larger test volume.

At the time of writing this book, one 1-m^3 chamber is in commercial operation in North America with perhaps a half dozen or so in Europe (and maybe that many again used for private test purposes). The number of 20-L chambers in operation globally is likely over 100. The reasons for this order of magnitude difference are related to cost-effectiveness; 20-L testing is intrinsically more economical than 1-m³ testing in terms of capital, operating, and sample costs.

The matter of overdriving effects on marginally explosible dusts is therefore important and needs to be resolved. There is a clear call from industry for cost-effective risk reduction measures appropriate for the actual degree of hazard. There are also clear incentives for test facility operators as they strive to provide reliable, unambiguous data to their clients.

19.4 REALITY

As the discussion in this chapter has demonstrated, commonly determined dust explosion parameters are not fundamental material properties. Any lingering doubt can be removed by reference to the American Society for Testing and Materials (ASTM) test standards for determining most of the parameters identified in Table 5.1. All contain wording identical or similar to the standard [14] for determination of P_{max} and K_{St} (p. 3): "The values obtained by this testing technique are specific to the sample tested and the method used and are not to be considered intrinsic material constants."

Some parameters such as P_{max} and K_{St} have specific end uses for the test data generated—e.g., the previously mentioned design of explosion mitigation systems. All parameters can provide a relative comparison of explosibility from at least two perspectives. First, a comparison can be made between the relative explosibility of a given material in various formulations with respect to particle size distribution or moisture content (expressed, for example, by ease of ignition as reflected in different MIE values). Second, a comparison can be made for materials under different processing scenarios. For example, it would likely not be practical to operate at concentrations less than the minimum explosible value for a dust having an MEC of, say, 40 g/m³. This preventive strategy may be feasible, however, for a material that has an MEC of the order of, say, 1000 g/m³. The concept of a risk-based approach is expanded upon in the next chapter.

19.5 WHAT DO *YOU* THINK?

The data in Table 19.2 were determined by testing with a Siwek 20-L chamber according to ASTM E1226 [14]. P_m is the maximum value of overpressure occurring in a given test at a specified dust concentration; similarly $(dP/dt)_m$ is the maximum value of rate of pressure rise occurring in a given test at a specified dust concentration. These parameters can be best understood by referring to Figure 11.8(a). Here, one can clearly see a peak overpressure that corresponds to P_m after correction for the ignitor pressure contribution and vessel cooling. The slope of the tangent drawn to the steepest rising portion of the pressure/time curve between approximately 120 and 130 ms would yield $(dP/dt)_m$. [From basic calculus, (dP/dt) is the time rate of change of pressure; the subscript m indicates the maximum value for a particular pressure/time trace.]

TABLE 19.2 P_{max}/K_{St} Data Determined in a Siwek 20-L Chamber According to ASTM E1226 [14]

Dust concentration [g/m³]	P_m [bar(g)]	$(dP/dt)_m$ [bar/s]
Series 1		
125	7.4	476
250	8.8	764
500	8.1	851
750	7.2	583
1000	6.9	476
Series 2		
125	7.0	611
250	8.7	932
500	7.8	970
750	6.9	647
Series 3		
125	6.9	493
250	8.3	916
500	7.7	753

In addition to other test protocol features, following ASTM E1226 [14] means using standardized values of dispersing air pressure (20 bar(g)), ignition delay time (60 ms), and ignition energy (10 kJ by means of two 5-kJ ignitors). The dust concentrations for testing are also established in the protocol, with progressive increments of 250 g/m³ being used. If testing is required at concentrations less than 250 g/m³, a halving procedure is followed (125 g/m³, 60 g/m³, etc.).

Three test series are conducted. The first series covers a range of dust concentrations with the objective of determining the concentration(s) at which the peak values of P_m and $(dP/dt)_m$ occur within the series. These peak values of the two parameters do not necessarily occur at the same dust concentration within a given test series. A second series is then conducted at, and on either side of, the maxima concentration(s). A similar third test series is undertaken. The concentrations at which the peak values of P_m and $(dP/dt)_m$ occur from series to series will not necessarily be the same (although they should not typically vary by more than one concentration interval). All of these features are evident in Table 19.2. Do you see them?

P_{max} is calculated by averaging the highest values of P_m from each of the three test series. Similarly, $(dP/dt)_{max}$ is calculated by averaging the highest values of $(dP/dt)_m$ from each of the three test series. K_{St} is then calculated from $(dP/dt)_{max}$ by means of Equation 19.1, with the vessel volume (V) in this case being 20 L or 0.02 m^3.

As an exercise, perform the calculations for P_{max} and K_{St} from the data in Table 19.2. This will help to demystify these parameters—especially if you have never done this sort of calculation before. You should get the following values: P_{max} = 8.6 bar(g), $(dP/dt)_{max}$ = 912 bar/s, and K_{St} = 248 bar·m/s.

There are more aspects to P_{max}/K_{St} determination than this one example can cover. Many other experimental considerations are important—such as the tolerance on series-to-series data repeatability, quality assurance of the oxygen concentration in compressed air sources, equipment calibration requirements, and the relatively high maxima concentrations for some materials such as fine iron powder. Further, and as previously mentioned in this book, global standards other than ASTM E1226 [14] do exist for P_{max}/K_{St} measurement; vessels having volumes greater than 20 L are also used for this purpose.

Established dust explosion test facilities understand these features and recognize the importance of their consideration in generating industrially relevant K_{St} data via standardized testing using standardized equipment. Anything less is simply a maximum rate of pressure rise multiplied by the cube root of a vessel volume. It is not a K_{St}. (See Section 19.2.)

REFERENCES

[1] Agarwal A. Dust explosions: prevention and protection. Chemical Engineering November 2012:26–30.

[2] Buffam J, Cox K, Schies H. Measurement of laminar burning velocity of methane-air mixtures using a slot and Bunsen burner. Worcester, MA: Bachelor of Science Major Qualifying Project, Worcester Polytechnic Institute; 2008.

[3] Goroshin S, Tang F-D, Higgins AJ, Lee JHS. Laminar dust flames in a reduced-gravity environment. Acta Astronautica 2011;68:656–66.

[4] Skjold T. Review of the DESC project. Journal of Loss Prevention in the Process Industries 2007;20:291–302.

[5] Bind VK, Roy S, Rajagopal C. A reaction engineering approach to modeling dust explosions. Chemical Engineering Journal 2012;207–208:625–34.

[6] Zalosh R. Explosion venting data and modeling research project. Literature review. Quincy, MA: The Fire Protection Research Foundation; 2008.

[7] Eckhoff RK. Dust explosions in the process industries, 3rd ed. Boston, MA: Gulf Professional Publishing/Elsevier; 2003.

[8] Kuhner. Final report. Calibration-round-robin. CaRo 11. Birsfelden, Switzerland: Adolf Kuhner AG; 2011. (Available at: http://safety.kuhner.com/tl_files/kuhner/product/safety/PDF/B052_256CaRo11.pdf; last accessed November 26, 2012.)

[9] Barton J, editor. Dust explosion prevention and protection. A practical guide. Rugby, UK: Institution of Chemical Engineers; 2002.

[10] Dastidar AG, Amyotte PR, Pegg MJ. Factors influencing the suppression of coal dust explosions. Fuel 1997;76:663–70.

[11]   Amyotte PR. Solid inertants and their use in dust explosion prevention and mitigation. Journal of Loss Prevention in the Process Industries 2006;19:161–73.

[12]   Cloney CT, Ripley RC, Amyotte PR, Khan FI. Quantifying the effect of strong ignition sources on particle preconditioning and distribution in the 20-L chamber. Krakow, Poland: Proceedings of Ninth International Symposium on Hazards, Prevention, and Mitigation of Industrial Explosions; July 22–27, 2012.

[13]   Bucher J, Ibarreta A, Marr K, Myers T. Testing of marginally explosible dusts: evaluation of overdriving and realistic ignition sources in process facilities. College Station, TX: Proceedings of 15th Annual Symposium, Mary Kay O'Connor Process Safety Center, Texas A&M University; October 23–25, 2012, pp. 688–97.

[14]   ASTM. ASTM E1226–10, Standard test method for explosibility of dust clouds. West Conshohocken, PA: American Society for Testing and Materials; 2010.

Chapter 20

Myth No. 19 (Pentagon): It Makes Sense to Combine Explosion Parameters in a Single Index

Earlier discussion in Chapter 5 indicated that the first three parameters in Table 5.1 [P_{max}, $(dP/dt)_{max}$, and K_{St}] address the *consequence severity* component of risk; the remainder (MEC, MIE, MIT, LIT, and LOC) are relevant to the issue of *likelihood of occurrence*. With this division of focus area, it might seem beneficial to combine parameters dealing with *both* risk components into a single measure of hazard potential.

There are certainly precedents for this approach in the field of process safety. For example, the Dow Fire & Explosion Index (F&EI) methodology uses a series of penalty factors to quantify process hazards related to both magnitude and probability of loss from fires and explosions [1]. The end result of the calculation procedure is a single number—the F&EI—that is useful both as a tool for relative risk ranking and as an indication of whether further risk review is warranted for the facility in question. The Dow Chemical Exposure Index (CEI) provides a similar avenue for addressing process hazards related to the release of toxic materials [2].

For purposes of identifying dust explosion hazards and assessing dust explosion risks, however, the combining of disparate explosibility parameters into a single index is fraught with theoretical weaknesses and practical uncertainties. As we explore in this chapter, it is the individual parameters themselves that should be considered in a coordinated fashion to manage the risks of dust explosions.

20.1 USBM INDICES

In the early 1960s, the U.S. Bureau of Mines, or USBM (now part of NIOSH, the U.S. National Institute for Occupational Safety and Health, as explained in Section 13.2), developed the following expressions:

$$Ignition\ Sensitivity = \frac{[MIT \cdot MIE \cdot MEC]Pittsburgh\ Coal\ Dust}{[MIT \cdot MIE \cdot MEC]Sample\ Dust} \quad (20.1)$$

$$Explosion\ Severity = \frac{[P_{max} \cdot (dP/dt)_{max}]Sample\ Dust}{[P_{max} \cdot (dP/dt)_{max}]Pittsburgh\ Coal\ Dust} \quad (20.2)$$

$$Explosibility\ Index = Ignition\ Sensitivity \cdot Explosion\ Severity \quad (20.3)$$

(See Section 6.1 for the size and volatility characteristics of the Pittsburgh coal dust referred to in Equations 20.1 and 20.2.)

By the late 1980s, Equations 20.1–20.3 were termed *outdated* by the very institution that had originally proposed them [3]. Yet to this day, these expressions—particularly ignition sensitivity and explosion severity as determined by Equations 20.1 and 20.2, respectively—continue to be referenced in some jurisdictions. In North America, Equations 20.1 and 20.2 find widespread use in various recommended practices, guidelines, codes, and regulations (largely for the purpose of area classification for electrical equipment).

There is, however, little scientific or engineering rationale for the existence of Equations 20.1–20.3 given that their development was not based on detailed theoretical or accident statistic analyses [3]. The combination of ignition sensitivity and explosion severity via Equation 20.3 is particularly problematic in yielding a dust explosion hazard rated as *weak*, *moderate*, *strong*, or *severe* depending on the magnitude of the explosibility index [3]. It must be recognized that virtually *any* dust explosion hazard has the potential to cause injury and property damage regardless of the subjective, qualitative rating it might be assigned by an indexing methodology. Similarly, although the use of K_{St} values to classify dusts as St 1 (1–200 bar·m/s), St 2 (201–300 bar·m/s), or St 3 (> 300 bar·m/s) continues to have some relevance, qualitative explosion descriptors of *weak* (St 1), *strong* (St 2), and *very strong* (St 3) should be avoided.

Martin Hertzberg, then of the U.S. Bureau of Mines, wrote [3] (pp. 2–3)

...the consensus is that the critical parameter in obtaining a hazard evaluation is not the explosivity properties of a dust relative to some other dust (such as Pittsburgh seam coal), but rather, their values relative to the operating conditions within the industrial system being evaluated.

Hertzberg [3] thus makes the same point as this book's recurrent theme of adopting a risk-based approach to alleviating the dust explosion problem.

As a practical example, the testing protocol used for new materials by one company involves an initial determination of minimum ignition energy (MIE). P_{max}/K_{St} testing may be further commissioned depending on the value measured for MIE. The rationale behind this approach is that if MIE determination indicates the potential for ignition by typical electrostatic discharges, then venting will be undertaken as a mitigation measure (hence necessitating further testing for P_{max}/K_{St}). If, on the other hand, electrostatic ignition is not deemed probable, then ignition source control is relied on as a primary preventive measure (among others) without the introduction of explosion relief vents. Implicit in this approach is an assessment and acceptance of risk.

There are, of course, occasions when examination of a dust's explosibility parameters relative to those for another material is entirely warranted. Consider, for example, the paper by Caine and Ackroyd [4] in which they explain their company's growth of its agribusiness by acquisition of seed processing facilities. With established expertise in active ingredient (AI) manufacturing, a comparison of AI explosibility with the parameters for corn dust (see Table 20.1) seems prudent. There is thus a clear rationale for concluding that corn processing dust is less ignition sensitive than typical AIs, although similar consequences would result if ignited [4]. This comparison would be of little significance if not for the fact that the company processes both bulk seeds and biologically active powders. In accordance with the earlier quote from the critique by Hertzberg [3], consideration of dust explosion hazards relative to operating conditions is then feasible.

TABLE 20.1 Selected Explosion Data for Corn Dust

P_{max} [bar(g)]	K_{St} [bar·m/s]	MIE [mJ]	MIT [°C]
6–8	100	> 30	> 400

(adapted from Caine and Ackroyd [4])

TABLE 20.2 Selected Explosion Data for Coal and Biomass Dusts

Dust	Ignition sensitivity			Explosion severity	
	LIT [°C]	MIT [°C]	MIE [mJ]	P_{max} [bar(g)]	K_{St} [bar·m/s]
Coal (< 63 µm)	> 400	670	> 1000	7.0	81
Wood (< 63 µm)	330	420	50–60	8.2	161
Dried sewage sludge (< 63 µm)	280	460	100–500	7.6	143

(adapted from Merritt [5])

A second example of this type can be found in the work of Merritt [5], who contrasts coal dust explosibility with that of biomass resources including wood and dried sewage sludge (see Table 20.2). One might wonder why there is a need for such a comparison; the reasoning becomes apparent when examining the case study of a company planning to use an available coal conveyor system to transfer pelletized biomass [5]. As indicated by the data in Table 20.2, biomass dust explosions generally result in more severe consequences (P_{max} and K_{St}) than coal dust, and the biomass dusts are more prone to hot-surface ignition (LIT and MIT) and spark ignition (MIE) than is coal dust. This is also a further example of the requirement for effective management of change as detailed in Section 11.3.

Two observations relevant to the current discussion can be made with respect to Table 20.2. First, the data provided are for samples of the different dusts having essentially the same particle size distribution (at least in terms of the upper limit). Second, the terms *ignition sensitivity* and *explosion severity* are used in Table 20.2. There is, however, nothing inappropriate about this usage; the individual explosion parameters are kept separate and distinct with no attempt to group them in a single value as per Equations 20.1 and 20.2.

Finally, reference is made to the article by Franck [6] on the sawmill dust explosions that occurred in the western Canadian province of British

Columbia in January and April of 2012. Here, we see recognition of the need, at times, for comparison of a given dust to a particulate form of the same material under different conditions. In the case of the British Columbia sawmill explosions (the causes of which have not been definitively identified at the time of writing this book), this means examination of whether dust from beetle-infested wood is somehow more explosible than non-infested wood dust. (Approximately 70% of the wood supply to the two sawmills in question was "beetle-kill"—wood that had been infested with mountain pine beetles [6].)

If this is shown to be the case, it would have to be due to some fundamental cause related to dust reactivity (particle size, moisture content, chemical composition, etc.). The concluding sentence in Franck's paper [6] (p. 17) again provides evidence of the importance of management of change:

If beetle-killed wood is indeed found to have been a factor, this will highlight the need for constant vigilance in situations where the nature of combustible dusts within a processing environment changes over time.

20.2 ASSESSMENT AND MANAGEMENT OF DUST EXPLOSION RISKS

In his 2009 review of the status and developments in basic knowledge and practical application of dust explosion prevention and mitigation measures, Eckhoff [7] gives "six perspectives for the future":

- Tailored and differentiated approaches that lead to more cost-effective safety measures
- CFD (computational fluid dynamics) modeling with experimental validation
- Evolution of the field of ignition source evaluation to include, for example, mathematical modeling
- Combined protection solutions (e.g., venting and automatic suppression)
- Inherently safer process design, which has as a basic requirement the knowledge and systematic use of powder science and technology
- high-quality training and education.

A unifying thread running through these six topics is the notion of addressing the dust explosion problem via assessment and management of risk. Development of improved risk assessment measures is also a key factor in strengthening the three pillars Knegtering and Pasman [8] have proposed as a basis for process safety improvements in the 21st century:

- An adequate process safety measurement system
- An effective and efficient evaluation and analysis of the observations made, accompanied by a continuous learning system
- A holistic approach to process safety management and control.

Relevant examples in the current book include: (i) a general discussion of *hazard* and *risk* (Section 5.1), (ii) reference to computational models such as Dust Explosion Simulation Code (DESC), which are key elements of quantitative consequence analysis (Section 6.5), and (iii) introduction of likelihood and outcome analysis tools such as fault trees, event trees, and bow-tie diagrams (Section 9.4).

Coupled with the need identified here to consider dust explosions as process incidents, it is heartening to see the recent efforts aimed at application of classical risk assessment and management techniques to this field. One sees, for example, the use of several qualitative methodologies by Marmo et al. [9] in their analysis of aluminum dust explosion risk:

- Checklist analysis (CA)
- What-if analysis (WI)
- Failure modes and effects analysis (FMEA)
- Hazard and operability analysis (HAZOP).

The conceptual framework for dust explosion prevention and mitigation does appear to be advancing from an emphasis on *hazards* (with an accompanying reliance on primarily engineered safety features) to a focus on *risk* (with an accompanying reliance on hierarchical, risk-based, decision-making tools). An initiative from TNO, Netherlands [10], illustrates an interesting approach to incorporating dust explosion likelihood and consequence severity considerations into a quantitative risk assessment tool. This work by van der Voort et al. [10] is one of the first such uses of quantitative risk assessment (QRA) reported in the open literature.

An overview of their tool is given in Figure 20.1. Consideration of what the authors have called "external safety" can be seen in their choice of explosion effects—debris and fragment throw (missiles), flame effects (thermal fluxes), blast (overpressures), and bulk outflow (material effects). This comprehensive suite of consequences was motivated by the 1997 grain storage facility explosion in Blaye, France (Section 12.1), in which most of the 11 fatalities were caused by debris fragments and grain outflow [10].

Abuswer et al. [11] also propose a quantitative risk management framework for dust and hybrid mixture explosions. Figure 20.2 shows their use of fault tree analysis (FTA) to demonstrate—qualitatively, in this case—the basic criteria of the explosion pentagon. Of interest is the appearance in Figure 20.2 of each of the relevant explosion likelihood parameters in Table 5.1 (MEC, MIE, MIT, and MOC). (MOC here refers to minimum oxygen concentration, a synonym for LOC.) There are, of course, different ways of representing the explosion pentagon in a fault tree; you may wish to compare Figure 20.2 with the dust explosion fault tree you developed in Section 9.4.

In addition to demonstrating the application of a CFD code to the task of dust explosion consequence analysis, Davis et al. [12] reinforce the need for risk-based decision making. These authors show that determination of an acceptable level of risk can be facilitated by means of a risk matrix like the five-by-five

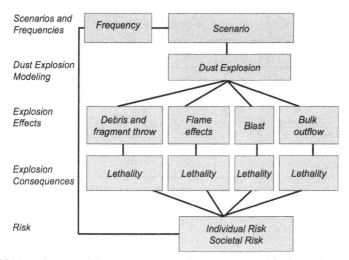

FIGURE 20.1 Overview of the QRA (quantitative risk assessment) tool for dust explosions developed by TNO, Netherlands [10].

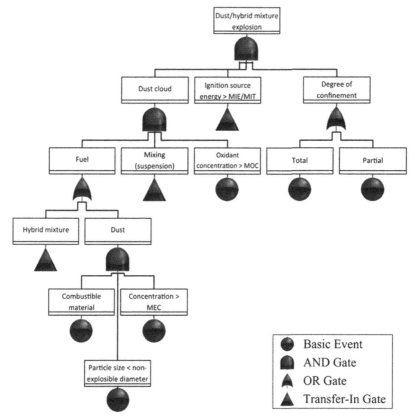

FIGURE 20.2 Fault tree representation of the explosion pentagon for dust and hybrid mixture explosions [11].

V	C	B	A	A	A
IV	D	C	B	A	A
III	E	D	C	B	A
II	E	E	D	C	B
I	E	E	E	D	C
	I	**II**	**III**	**IV**	**V**
	Probability				

(left vertical axis label: Consequence)

FIGURE 20.3	Sample risk matrix for dust explosions [12].

matrix given in Figure 20.3. Here, the probability and consequence categories range from *I* to *V* in order of increasing likelihood and severity, respectively. Risk levels are from *A* to *E*; *A* denotes a very high and unacceptable level of risk for which risk-reducing measures must be implemented, and *E* corresponds to a very low and acceptable level of risk for which risk-reducing measures are not required [12]. This way of thinking is not unlike the ALARP principle referenced in Section 11.1. (The original paper by Davis et al. [12] may be consulted for a full description of each probability category, consequence category, and risk level.)

As a final example in this section, the interesting work of Vignes et al. [13] is cited. These authors provide a strategic approach for assessing the risk of aluminum nano-powder explosions by adopting the concept of a *risk score* based on three parameters: (i) explosion consequences, (ii) explosion frequency, and (iii) level of confidence in prevention and mitigation barriers [13]. Of note, their method is specifically based on the results of a previous study in which they measured the explosibility parameters (P_{max}, K_{St}, MEC, MIE, and MIT) of nano-size aluminum [14].

20.3 MATERIAL SAFETY DATA SHEETS

It is therefore critical that explosibility data be provided to plant workers in the form of the individual parameters listed in Table 5.1, not as a combination of these parameters in an explosibility index. This can be accomplished via material safety data sheets (MSDSs), which are an essential ingredient for ensuring a safer industrial workplace. MSDSs provide critical information on the physical properties of a material as well as key indicators of the hazardous nature of the material. As such, they play an important role in fulfilling a worker's right to know about facility hazards and risks (Section 18.3) as part of the system known as the Workplace Hazardous Materials Information System (WHMIS).

As described by Amyotte et al. [15], an important development in this regard is the analysis of 140 material safety data sheets conducted by the U.S. Chemical Safety Board [16]. The CSB determined that MSDSs for combustible dusts do not clearly and consistently communicate dust explosion hazards because of their inadequate identification of dust explosibility parameters and the hazards that may be expected to arise through material handling and processing.

Amyotte et al. [15] performed a similar analysis for a single material—polyethylene. A simple procedure was deliberately chosen to simulate the action a plant worker might take in the absence of both an MSDS and knowledge of the particular manufacturer. An Internet search for *Polyethylene MSDS* yielded the expected multiple hits. The first 12 MSDSs were selected for analysis (12 being a number someone might choose to investigate before deciding that as much information as was available had been retrieved).

The results are shown in Table 20.3. Findings are in general agreement with the CSB study [16], especially with respect to physical data on dust explosibility. None of the 140 MSDSs reviewed by the CSB [16] and none of the 12 MSDSs in Table 20.3 contain any explosibility data for the dust in question. This is clearly a reality that needs urgent attention so as not to perpetuate the myths described in the current book.

20.4 REALITY

The individual explosibility parameters determined for a given combustible dust represent the factors necessary for hazard analysis and risk assessment. MEC, MIE, MIT, LIT, and LOC help to determine the likelihood of occurrence, and P_{max} and K_{St} are measures of consequence severity. These parameters find optimal effectiveness when considered both separately and in relation to one another—not when combined in an arbitrary manner to generate a single index of presumed significance. Unambiguous communication of hazardous features of combustible dusts by means of traditional avenues such as material safety data sheets is currently problematic.

Assessment of dust explosion risks and management of residual risk following the implementation of risk-reducing measures require consideration of both material hazards and the conditions under which these same materials are processed. Both qualitative and quantitative risk assessment techniques are important in this regard. These methodologies have been used to address other process safety issues including flash fires, vapor cloud explosions, and dispersion of toxic materials. Their application in the dust explosion field is strongly encouraged.

20.5 WHAT DO *YOU* THINK?

There is clearly a problem with material safety data sheets (MSDSs) and the information they convey on dust explosibility parameters and general dust

TABLE 20.3 Review of Polyethylene MSDSs by Amyotte et al. [15] According to Criteria Recommended by the U.S. Chemical Safety Board [16]

MSDS No.	Hazard stated explicitly?	In hazard information section?	Hazard warning repeated?	Dust explosibility data?	Reference to NFPA standard?	Warning against accumulation?
1	YES	YES	YES	NO	YES	YES
2	NO	NO	NO	NO	NO	NO
3	NO	NO	NO	NO	NO	NO
4	YES	YES	YES	NO	NO	YES
5	YES	YES	YES	NO	NO	YES
6	YES	YES	YES	NO	YES	YES
7	YES	NO	NO	NO	YES	YES
8	YES	YES	YES	NO	NO	YES
9	NO	NO	NO	NO	NO	NO
10	YES	YES	YES	NO	YES	NO
11	NO	NO	NO	NO	NO	NO
12	NO	NO	NO	NO	NO	NO

explosion hazards. In response, official calls have been made in North America to enhance both the quality and quantity of combustible dust hazard information contained in MSDSs [17,18].

As another example of the need for such improvements, Ebadat and Prugh [19] report on the investigation following an explosion in a facility pulverizing aluminum-alloy ingots. It was concluded that neither the available MSDS nor employee training adequately addressed the explosibility of aluminum and aluminum–alloy dusts [19].

But are material safety data sheets really the best vehicle to achieve the goal of providing comprehensive warnings of the explosible nature of industrial dusts? What is your opinion on this point?

Are the MSDSs for combustible dusts handled in your facility adequate in terms of explosion hazard data and warnings? Is there a different or better way to provide this information to the front-line operators, supervisors, and managers whose lives may depend on the needed knowledge?

There appear to be two schools of thought on this matter. Some professionals feel that MSDS suppliers must provide the requisite data—either voluntarily or as a regulated requirement [20]. Others see little value in this approach given the influence of site-specific conditions on dust explosion behavior [20]. Again, how do you feel about MSDSs and their ability to convey dust explosion information?

As for me, I feel we should at least stop assuming that MSDSs will tell us everything we need to know about the hazardous nature of a given material. As a provider of dust explosibility data to industry, I require that an MSDS be supplied with every sample I receive for testing. I am most interested in any unique hazards related to, for example, toxicological properties or material incompatibility—i.e., issues of concern from a testing perspective. I am not looking for explosibility data or explosion warnings in the provided MSDSs, so I am not disappointed when they invariably contain little, if any, information of this nature. (Arguably, this is not entirely surprising given that the purpose of explosibility testing is to determine such data.)

There are, of course, notable exceptions. The MSDS accompanying the CaRo 11 niacin dust sample used in calibration-round-robin testing (as described in Section 19.2) contains ample dust explosion warnings as well as values for most of the parameters listed in Table 5.1. (One need not worry about the validity of the round-robin testing; the complete data sets for the parameters determined are submitted to a third party for independent assessment and verification.)

In preparing this chapter, I decided to repeat on a smaller scale the exercise that led to the development of Table 20.3 for polyethylene. I chose zinc stearate as the sample dust because we had tested this material in our laboratory, and I knew it to have significant explosion consequences, as shown in Table 20.4. I then did a quick Internet search for *Zinc Stearate MSDS* using a standard search engine. The results for the first six hits are given in Table 20.5 according to four key MSDS sections.

TABLE 20.4 Selected Explosion Data for Zinc Stearate Dust

Data source	Median particle diameter [μm]	P_{max} [bar(g)]	K_{St} [bar·m/s]	MEC [g/m³]	MIE [mJ]	MIT [°C]
IFA database [21]	<10	9.2	286	30	< 5	380
Standardized 20-L testing	< 45	8.9	281	Not measured	Not measured	Not measured
Standardized 20-L testing	< 45	8.3	213	Not measured	Not measured	Not measured

My overall conclusion from comparing Tables 20.4 and 20.5 is that the selected zinc stearate MSDSs are woefully inadequate in depicting the true picture of zinc stearate explosion likelihood and consequence severity. Granted, Table 20.4 contains particle size references that are not commonly found in an MSDS; further, some entries in Table 20.5 do give qualitative indicators of the explosion potential of zinc stearate dust. Nevertheless, the limited quantitative data given in Table 20.5 are generally misleading.

I am speaking of MSDS Nos. 2 and 5 in Table 20.5. With respect to MSDS No. 2, a minimum explosible concentration of 0.02 g/L (20 g/m³) is reasonable as per Table 20.4. However, a value of 68 lb_f/in^2 for P_{max} corresponds to 4.7 bar and is well below all the values for P_{max} in Table 20.4 (whether it is 4.7 bar absolute or gauge, which is not known). This is not unexpected because the reference for the data in MSDS No. 2 is given as *Bureau of Mines, 1968*. Explosibility data determined by the U.S. Bureau of Mines in 1968 would undoubtedly have been acquired through testing with a 1.2-L cylindrical vessel known as the Hartmann bomb. This apparatus is known to underestimate P_{max} and K_{St} while in many cases yielding realistic values of MEC.

I wonder how many people in need of accurate information are that familiar with the history of dust explosion testing. Is it a reasonable expectation that they should be?

As for MSDS No. 5, the lower explosion limit can be taken to mean the minimum explosible concentration. An MEC of 0.22 g/L corresponds to a value of 220 g/m³, which seems high by any measure (Table 20.4 or otherwise). A typographical error, perhaps?

TABLE 20.5 Review of Zinc Stearate MSDSs for Dust Explosion Information According to MSDS Section

MSDS No.	Hazards identification	Fire and explosion data	Physical and chemical properties	Stability and reactivity data
1	None given	"Risks of explosion of the product in presence of mechanical impact: Not available." "Risks of explosion of the product in presence of static discharge: Not available." "Special remarks in explosion hazards: Not available."	None given	None given
2	*Warning! May form combustible dust concentrations in air*	"Explosion: Fine dust dispersed in air in sufficient concentrations, and in the presence of an ignition source is a potential dust explosion hazard. Minimum explosible concentration: 0.02 g/L (air) (Bureau of Mines, 1968). Maximum explosion pressure: 68 lbf/in2 at 0.3 oz/ft3. Sensitive to static discharge." "Special information: Explosion hazards apply only to dusts, not to granular forms of this product."	None given	*Conditions to avoid: Heat, flames, ignition sources.*
3	None given	None given	None given	None given
4	None given	"Special fire fighting procedures: Slight fire hazard when exposed to heat or flame—any finely dispersed particles are explosive. Avoid dust concentrations. Dust-air mixtures could cause explosions."	None given	*Conditions to avoid: May burn but does not readily ignite.*
5	None given	"General information: Dusts at sufficient concentrations can form explosive mixtures with air." "Explosion limits: Lower—0.22 g/L; Upper—Not available."	None given	None given
6	None given	"Unusual fire and explosion hazards: Dust explosions can occur under conditions of high dust concentration in the presence of an ignition source."	None given	None given

So the question remains. Are MSDSs (or SDSs – safety data sheets) the best way of achieving the goal of providing comprehensive warnings of the explosible nature of industrial dusts? If you have an alternative—or a way to make MSDSs more amenable to dust hazard communication—I encourage you to state your case. One approach is via the archival literature or refereed conference proceedings. Additionally, I would be most interested to hear your thoughts on the matter; my email address is *paul.amyotte@dal.ca* (as of January 25, 2013).

REFERENCES

[1] Dow. Fire & explosion index hazard classification guide, 7th ed. New York, NY: American Institute of Chemical Engineers; 1994.

[2] Dow. Chemical exposure index guide, 2nd ed. New York, NY: American Institute of Chemical Engineers; 1993.

[3] Hertzberg M. A critique of the dust explosibility index: an alternative for estimating explosion probabilities. U.S. Bureau of Mines Report of Investigations, RI 9095. Pittsburgh, PA: U.S. Bureau of Mines; 1987.

[4] Caine PJ, Ackroyd GP. Exploding Corn. Southport, UK: Hazards XXIII, IChemE Symposium Series No. 158; November 13–15, 2012, pp. 71–6.

[5] Merritt M. Biomass: energy remedy or safety headache? The Chemical Engineer 2012;857:30–4.

[6] Franck A. Sawmill dust explosions in Canada: under the beetle's shadow. Hazardous Area International 2012;6(3):15–17.

[7] Eckhoff RK. Dust explosion prevention and mitigation. Status and developments in basic knowledge and in practical application. International Journal of Chemical Engineering 2009; Article ID 569825, 12 pp.

[8] Knegtering B, Pasman H. Safety of the process industries in the 21st century: a changing need of process safety management for a changing industry. Journal of Loss Prevention in the Process Industries 2009;22:162–8.

[9] Marmo L, Cavallero D, Debernardi ML. Aluminum dust explosion risk analysis in metal workings. Journal of Loss Prevention in the Process Industries 2004;17:449–65.

[10] van der Voort MM, Klein AJJ, de Maaijer M, van den Berg AC, van Deursen JR, Versloot NHA. A quantitative risk assessment tool for the external safety of industrial plants with a dust explosion hazard. Journal of Loss Prevention in the Process Industries 2007;20:375–86.

[11] Abuswer M, Amyotte P, Khan F. A quantitative risk management framework for dust and hybrid mixture explosions. Journal of Loss Prevention in the Process Industries 2013;26: 283–9.

[12] Davis SG, Hinze PC, Hansen OR, van Wingerden K. Does your facility have a dust problem: methods for evaluating dust explosion hazards. Journal of Loss Prevention in the Process Industries 2011;24:837–46.

[13] Vignes A, Munoz F, Bouillard J, Dufaud O, Perrin L, Laurent A, Thomas D. Risk assessment of the ignitability and explosivity of aluminum nanopowders. Process Safety and Environmental Protection 2012;90:304–10.

[14] Bouillard J, Vignes A, Dufaud O, Perrin L, Thomas D. Ignition and explosion risks of nanopowders. Journal of Hazardous Materials 2010;181:873–80.

[15] Amyotte P, Lindsay M, Khan F. The role of material safety data sheets in dust explosion prevention and mitigation. Bruges, Belgium: Proceedings of 13th International Symposium on Loss Prevention and Safety Promotion in the Process Industries, vol 2. June 6–9, 2010, pp. 427–30.

[16] CSB. Investigation report—combustible dust hazard study. Report No. 2006-H-1. Washington, DC: U.S. Chemical Safety and Hazard Investigation Board; 2006.

[17] Health Canada. Hazard-specific issues—dust explosibility. (Available at: http://www.hc-sc. gc.ca/ewh-semt/occup-travail/whmis-simdut/dust-poussiere-eng.php; last accessed December 9, 2012.)

[18] OSHA. Hazard communication guidance for combustible dusts. OSHA 3371–08. Washington, DC: Occupational Safety and Health Administration, U.S. Department of Labor; 2009.

[19] Ebadat V, Prugh RW. Case study: aluminum-dust explosion. Process Safety Progress 2007;26:324–9.

[20] Shelley S. Preventing dust explosions: are you doing enough? Chemical Engineering Progress 2008;104(3):8–14.

[21] IFA. GESTIS-DUST-EX, Database: combustion and explosion characteristics of dusts. Sankt Augustin, Germany: Institut für Arbeitsschutz der Deutschen Gesetzlichen Unfallversicherung; 2012. (Available at: http://www.dguv.de/ifa/en/gestis/expl/index.jsp; last accessed December 5, 2012.)

Myth No. 20 (Pentagon): It Won't Happen to Me

And so we arrive at the final chapter presenting a specific myth and its corresponding reality. I feel that in some ways we have saved "the best for last"—or at least the most significant for last. For here we examine the philosophy that without the will to implement dust explosion risk reduction measures in a systematic and organized manner, all our technical solutions are for naught. Expressing this slightly more positively, we might say that the natural and applied sciences can learn much in this area from the management and social sciences. To this end, the discussion in the current chapter largely follows that given by Amyotte and Eckhoff [1] (with relevant excerpts).

The consequences of the belief that *it* (i.e., a dust explosion) *won't happen to me* are graphically illustrated by Figure 4.1. How else, other than an adherence to this myth, can one explain the Westray coal mine explosion? Referenced several times already in this book, Westray is the case study selected here for further examination from the perspectives of safety culture and a management system approach to safety.

First though, the next two sections deal with some fundamental features of these important concepts. I can think of no better rationale for considering them together than the title of Dennis Hendershot's recent paper: "Process Safety Management—You Can't Get It Right Without a Good Safety Culture [2]." A close second would be the following quotes (p. 17 and pp. 22–23, respectively) from the policy document titled "Process Safety Research Agenda for the 21st Century [3]":

To state that a situation is 100-percent safe has proven many times to be a gross exaggeration with serious consequences. All test methods and models have limitations. Compounding this issue are highly complex, individualized scenarios and a widely varying human fallibility factor.

and

The vapor cloud explosion at the isomerization unit of the BP Texas City refinery in March 2005 that resulted in 15 fatalities due to overfilling a column with hydrocarbons and subsequent discharge into the air triggered a thorough investigation, first by the U.S. Chemical Safety and Hazard Investigation Board and later by the BP U.S. Refineries independent safety review panel under the chairmanship of James A. Baker III, former Secretary of State. This investigation uncovered the management failures and lack of safety culture that led to the disaster. Other tremendously costly incidents as well as the Deepwater Horizon disaster have demonstrated the same trends. In hindsight, no new mechanisms or unknown hazards have been revealed. Unfortunately, knowledge about the risks involved has been available, but at the crucial moments of decision-making it is either not present, or it is ignored because of other pressures. At the very least, the decision made in absence of this information narrows the safety margin so that with a series of these kinds of decisions, the processes or operations reach a significantly higher level of risk.

21.1 SAFETY CULTURE

The fundamental issue of whether a company (in particular, its senior managers) believes it is possible to achieve a higher standard of safety—in essence whether a company believes safety is "the right thing to do"—has been addressed in a series of publications by sociologist Andrew Hopkins [4–7]. (See also Section 9.2.) I refer to Hopkins as a *sociologist* not as a label, but rather to differentiate his background and training from that of most technologists, applied scientists, and engineers working in the process and manufacturing industries. This is in keeping with my earlier remark that the field of dust explosions has much to learn from the social sciences.

Hopkins [5] describes three concepts that address a company's cultural approach to safety, and makes the argument that the three are essentially alternative ways of talking about the same phenomena: (i) safety culture, (ii) collective mindfulness, and (iii) risk-awareness. He further defines a safety culture as embodying the following subcultures: (i) a "reporting culture" in which people report errors, near-misses, and substandard conditions and practices; (ii) a "just culture" in which blame and punishment are reserved for behavior involving defiance, recklessness, or malice, such that incident reporting is not discouraged; (iii) a "learning culture" in which a company learns from its reported incidents, processes information in a conscientious manner, and makes changes accordingly; and (iv) a "flexible culture" in which decision-making processes are not so rigid that they cannot be varied according to the urgency of the decision and the expertise of the people involved [5].

Safety culture measurement requires the use of appropriate indicators. There has been considerable discussion recently in the literature on safety culture indicators (lagging and leading), and significant efforts aimed at developing suitable indicators have been undertaken by organizations such as the UK Health and Safety Executive (HSE) and the Center for Chemical Process Safety (CCPS) of the American Institute of Chemical Engineers (AIChE). An increasingly common thought being promoted by Hopkins and others [8] is that safety culture indicators must measure the effectiveness of the various measures comprising the risk control system. In other words, safety indicators must be related to the elements making up the safety management system being employed. (See Section 21.2.)

There is thus a strong link between the technical issues associated with dust explosion prevention and mitigation, and the collective beliefs or safety culture of an organization. This co-existence of technology and human action is clearly identified in the following passage [9] concerning the aftermath of the 2008 Imperial Sugar Company explosion [10]:

While the concepts of separation, isolation, and suppression form the core of Imperial's conspicuous changes in facility design and protection,...the more elusive aspect of its overall program remains tied to the human element, the most fragile variable in determining the prospects for success since it relies on the attitude of the workers. That's where [the company's] end game comes in: changing the fundamental culture at Imperial so that safety standards are not just tolerated, but genuinely valued.

Emphasis on cultural aspects of dust explosion prevention and mitigation continues at this particular facility, as evidenced by the following quote (p. 5) from the 2012 article by Franck [11]:

Sheptor [Imperial Sugar CEO John C. Sheptor] said experts believe a failed bearing overheated and touched off the explosions in the plant. Today, any accumulation of a 1/32 inch layer or more of sugar dust triggers a shutdown of the production line and will be investigated by a committee, he said. It had taken time to transform the culture so employees would take action themselves when such conditions were found, he added, but that has now been accomplished.

21.2 SAFETY MANAGEMENT SYSTEMS

Safety management systems are recognized and accepted worldwide as best-practice methods for managing risk. They typically consist of 10–20 program elements that must be effectively carried out to manage the risks in an acceptable way. This need is based on the understanding that once a risk is accepted, it does not go away; it is there waiting for an opportunity to happen unless the management system is actively monitoring company operations for concerns and taking preventive actions to correct potential problems. As indicated in Section 11.3, the cycle of *plan, do, check,* and *act* is the mantra of the management system approach to safety.

As a primary corporate objective, dust explosion prevention and mitigation fall within the scope of a process safety management system (i.e., a management system for process-related hazards such as fire, explosion, and release of toxic materials). Process safety management itself, or PSM, is defined as the application of management principles and systems to the identification, understanding, and control of process hazards to prevent process-related injuries and accidents.

A risk-based approach to PSM [12] consisting of 20 individual elements grouped into four thematic areas was identified in Section 11.3. The version of PSM [13] recommended by the Canadian Society for Chemical Engineering (CSChE) consists of the 12 elements shown in Table 21.1. Element 5 in Table 21.1—management of change—was previously discussed in Chapter 11.

In light of the treatment of inherently safer design (ISD) given in Chapter 15, a strong case can be made for the need to demonstrate a commitment to the principles of ISD within each of the PSM elements listed in Table 21.1 [14]. Thus, within element 4 (process risk management), the hierarchical arrangement of dust explosion prevention and mitigation measures shown in Table 15.2 would find direct application. Additionally, within element 8 (training and performance), a strong process safety culture would necessitate the provision of workplace training in dust explosion hazards. (See Section 18.3.) As illustrated by the analysis that follows for element 9 (incident investigation), similar observations can be made for the other PSM elements in Table 21.1.

Amyotte et al. [8] demonstrate that three of Hopkins' four safety subcultures [5] (*reporting, just,* and *learning*) have an explicit link to the PSM element of

TABLE 21.1 Twelve-Element PSM System of Canadian Society for Chemical Engineering [13]

No.	Element
1	Accountability: Objectives and Goals
2	Process Knowledge and Documentation
3	Capital Project Review and Design Procedures
4	Process Risk Management
5	Management of Change
6	Process and Equipment Integrity
7	Human Factors
8	Training and Performance
9	Incident Investigation
10	Company Standards, Codes and Regulations
11	Audits and Corrective Actions
12	Enhancement of Process Safety Knowledge

incident investigation. With respect to the subject of this book, a company's commitment to just and reporting cultures will ultimately be expressed in the number of combustible dust incidents (explosions *and* fires) that are reported. In the spirit of a leading indicator, it would be important to also measure the number of near-miss and at-risk behavior reports involving combustible dusts.

The outputs of reporting and investigation will manifest themselves as a measure of commitment to a learning culture; *lessons learned* is a key phrase in the industrial lexicon. The idea of learning from experience extends beyond the realm of incident investigation and into other safety management system elements (e.g., process risk management, management of change, and enhancement of process safety knowledge).

It may be tempting to dismiss this talk of safety culture and safety management systems as being unimportant in relation to the chemistry, physics, and engineering of dust explosions. Further, safety culture might be viewed by some as merely the current "hot topic" in industrial safety—i.e., a relatively new concept that, given time, will be replaced by something else.

On the first matter, the case study in the following section will, I hope, dispel any notion that well-established technical knowledge, in the absence of a strong safety culture and effective management system, is sufficient to ensure an acceptably low dust explosion risk. If there are lingering doubts, I encourage you to browse the Energy Institute's website and peruse its *Hearts and Minds* approach [15] to climbing the HSE (Health, Safety, and Environment) "culture ladder." There are convincing arguments here and in other programs that can be identified with an Internet search. (Recall the discussion in Section 8.3 on looking to authoritative, respected organizations for information on topics that may be outside one's comfort zone or field of expertise.)

On the second point—that safety culture is simply the "flavor of the month"—I offer the following quote from the final paragraph of Count Morozzo's report [16] describing the Turin flour warehouse explosion mentioned in Chapter 3:

Ignorance of the fore-mentioned circumstances, and a culpable negligence of those precautions which ought to be taken, have often caused more misfortunes and loss than the most contriving malice. It is therefore of great importance that these facts should be universally known, that public utility may reap from them every possible advantage.

The preceding passage makes an eloquent case for the importance of a strong safety culture, effective incident investigation, and the sharing of lessons learned. It is instructive to also note that it was written over two centuries ago.

21.3 WESTRAY COAL MINE EXPLOSION

John Thomas Bates, 56	Trevor Martin Jahn, 36
Larry Arthur Bell, 25	Laurence Elwyn James, 34
Bennie Joseph Benoit, 42	Eugene W. Johnson, 33
Wayne Michael Conway, 38	Stephen Paul Lilley, 40
Ferris Todd Dewan, 35	Michael Frederick MacKay, 38
Adonis J. Dollimont, 36	Angus Joseph MacNeil, 39
Robert Steven Doyle, 22	Glenn David Martin, 39
Remi Joseph Drolet, 38	Harry A. McCallum, 41
Roy Edward Feltmate, 33	Eric Earl McIsaac, 38
Charles Robert Fraser, 29	George S. James Munroe, 38
Myles Daniel Gillis, 32	Danny James Poplar, 39
John Philip Halloran, 33	Romeo Andrew Short, 35
Randolph Brian House, 27	Peter Francis Vickers, 38

Not just numbers, but ages. Not just numbers, but names. These are the 26 men killed on May 9, 1992, in the Westray coal mine at Plymouth, Nova Scotia, Canada [17]. These are the men whose names are forever carved in stone (see Figure 21.1).

The Westray disaster was introduced in Chapter 4. Methane levels in the mine were consistently higher than regulation because of inadequate ventilation design and operation. Dust accumulations also exceeded permissible levels due to inadequate cleanup of coal dust; additionally, there was no crew in charge of rock dusting (inerting the coal dust with limestone or dolomite). These and many other factors contributed to the poor work conditions that continually existed in the Westray mine and made it the site of an incident waiting to happen. All of these substandard conditions and practices could be attributed to the

FIGURE 21.1   Westray memorial with rays of light from a miner's lamp containing the names and ages of the 26 men killed in the Westray coal mine explosion [18]. The banner at the base of the memorial reads: NO MORE WESTRAYS. *(Photograph courtesy of United Steelworkers; Photograph by Peter Boyle.)*

lack of concern that management had toward safety issues in the mine and the absence of an effective safety management system [19].

The question may arise as to whether a coal dust explosion in an underground mine, which was initiated by a methane explosion, is a typical case in a dust explosion context. I suppose it would be considered quite typical if one believed the myth of dust explosions occurring only in coal mines and grain elevators. But as we know from Chapter 3, that is not one of the realities of dust explosions.

So again—why a coal mine explosion as an example here in our examination of cultural and management system aspects of process safety? There are, in fact, several reasons:

- First, the incident was the subject of a long-running public inquiry and is therefore well documented [17].
- Second, the explosion has been thoroughly analyzed from a root causation perspective [19].
- Third, all of the four major Canadian process-related incidents described in a recent article [18] on PSM were fires and/or explosions. Two of the incidents occurred in mines; two involved combustible dust. Westray—a coal dust explosion occurring in a mine—was one of the four incidents cited.

Lack of adequate loss prevention and management is generally due to deficiencies in one or more of three areas: (i) the safety management system itself, (ii) the standards identified and set for the safety management system, and (iii) the degree of compliance with such standards [20]. A particular management system may have missing components or may be entirely absent. Alternatively, the management system element may be present to some degree but could have inappropriate standards—or perhaps standards with which there is little or no compliance. Management system elements (Table 21.1) and their components that contributed to the Westray explosion include

- Management commitment and accountability to safety matters (which is a key element in establishing an effective company safety culture)
- Management of change procedures
- Incident investigation (including near-miss reporting and investigation)
- Training (orientation, safety, task-related, etc.)
- Task definition and safe work practices and procedures
- Workplace inspections and more proactive hazard identification methodologies
- Program evaluation and audits.

System standards that contributed to the Westray explosion include standards (i.e., level of performance) relating to virtually all the system elements and components in the preceding list, including inadequate

- Concern expressed by management toward safety matters (in terms of the standard of care one would reasonably expect and that, from a legal perspective, must be considered mandatory)
- Follow-through on inspections for substandard practices and conditions
- Action on hazard reports submitted by employees
- Job instructions for employees
- Equipment maintenance
- Scheduling of management/employee meetings to discuss safety concerns.

Compliance factors that contributed to the Westray explosion include

- Poor correlation between management actions and official company policy concerning the relationship between safety and production (as evidenced by the same management personnel holding responsibility for both production and underground safety)
- Inadequate compliance with industry practice and legislated standards and regulations concerning numerous aspects of coal mining (methane concentrations, rock dusting, control of ignition sources underground, etc.).

It is little wonder that Justice K. Peter Richard, the Westray Mine Public Inquiry Commissioner, titled his report "The Westray Story—A Predictable Path to Disaster [17]."

21.4 REALITY

The magazine *OHS Canada* posed the following question to six people who all worked in the health and safety field, but who had different educational backgrounds and professions: "Are all accidents preventable? [21]" Not surprisingly, six different answers were obtained. Two responses are particularly relevant to the current discussion [21]: (i) "No...but we should proceed on the basis that they are…" (from a university professor of occupational health and safety), and (ii) "The short answer is yes, you are required to plan and perform work as if all accidents were preventable" (from a lawyer).

The reality associated with the myth *it won't happen to me* is therefore not restricted to dust explosions and applies equally well to any process safety or occupational safety incident. Simply put: *What can happen, will happen if you believe it will not.*

Dust explosions must, however, be treated as process incidents—as was emphasized in the earlier discussion of the incident pyramid shown in Figure 18.1. A *process* safety culture and a *process* safety management system are needed to prevent and mitigate dust explosions. This is especially evident in the 2012 top ten most frequently cited workplace safety violations as reviewed by the U.S. Occupational Safety and Health Administration (in decreasing order) [22]: fall protection, hazard communication, scaffolding, respiratory protection, ladders, machine guarding, powered industrial trucks, electrical (wiring methods), lockout/tagout, and electrical (general requirements).

It is difficult to envisage how increased vigilance with respect to working at height (fall protection, scaffolding, and ladders) will help alleviate the dust explosion problem. This is an obvious and important concern for occupational safety; but that is not the subject of this book. Yes, better hazard communication will help in our efforts to reduce the impact of dust explosions—but only if the hazards communicated are those related to combustible dusts. Some of the electrical issues mentioned in the preceding paragraph undoubtedly relate to potential ignition sources. In reality, though, it is the material in Sections 21.1 and 21.2, and the final paragraph of Section 18.1, that should drive our dust explosion risk reduction efforts.

To conclude this final reality section related to a specific myth, I would like to quote (p. 108) from one of the reports [23] written about a recent tragedy in West Virginia that bears an eerie and disheartening similarity to the Westray disaster. The passage needs no further introduction:

The story of Upper Big Branch is a cautionary tale of hubris. A company that was a towering presence in the Appalachian coalfields operated its mines in a profoundly reckless manner, and 29 coal miners paid with their lives for the corporate risk-taking. The April 5, 2010 explosion was not something that happened out of the blue, an event that could not have been anticipated or prevented. It was, to the contrary, a completely predictable result for a company that ignored basic safety standards and put too much faith in its own mythology.

21.5 WHAT DO *YOU* THINK?

In his address to a 2012 meeting of the International Code Council, Dr. Rafael Moure-Eraso (Chair of the U.S. Chemical Safety Board) commented as follows [24]:

In 2010, as you know, the Hoeganaes powdered metals plant in Gallatin Tennessee had not just one, not just two, but three—three!—combustible dust flash fires. They killed a total of five workers and injured three others. Despite evidence released by the CSB and information that Hoeganaes had in its possession even before the first accident in January 2011, the company did not institute adequate dust control or housekeeping measures after the first and second accidents. Tennessee OSHA [Occupational Safety and Health Administration] and the City of Gallatin TN are monitoring the adequacy of the dust controls instituted by the company after the third explosion.

One might reasonably wonder why the company did not act to introduce these dust control measures until after the third incident. Perhaps a partial answer lies in Hopkins' description of the four subcultures of a *safety culture* [5], as summarized in Section 21.1, or in his explanation of *collective mindfulness*, which embodies the principle of *mindful organizing* and thus incorporates the following processes [5]:

- *Preoccupation with failure* so that a company is not lulled into a false sense of security by periods of success. A company that is preoccupied with failure would have a well-developed reporting culture.

- *Reluctance to simplify* data that may at face-value seem unimportant or irrelevant, but which may, in fact, contain the information needed to reduce the likelihood of a future surprise. (Note that simplification here is not a good thing, unlike the inherent safety principle of simplification described in Chapter 15.)

- *Sensitivity to operations* in which frontline operators and managers strive to remain as aware as possible of the current state of operations, and to understand the implications of the present situation for future functioning of the company.

- *Commitment to resilience* in which companies respond to errors or crises in a manner appropriate to deal with the difficulty, and a *deference to expertise* in which decisions are made by the people in the company hierarchy who have the most appropriate knowledge and ability to deal with the difficulty.

As a suggested exercise, you might want to choose one of these concepts—safety culture with its four subcultures [5] or collective mindfulness with its five processes [5]—and analyze the Hoeganaes incidents using the CSB case study report [25] as a basis.

Further insight into the need for corporate leadership in process safety culture and management can be gained by reading the Organization for Economic Co-operation and Development (OECD) report [26] on the subject. Quoting Bob Hansen, CEO of Dow Corning, in this report [26] (p. 12):

Creating a culture where all employees expect the unexpected and strive for error-free work is absolutely essential for success in process safety. This kind of culture is possible only through demonstrated leadership at all levels of the organization.

The notion that leadership and culture sit at the center of a continuous improvement cycle from risk-awareness through to action is illustrated by Figure 21.2. Table 21.2 gives a sample self-assessment question from the OECD report [26] for each of the five elements in Figure 21.2. Although intended for CEOs and

FIGURE 21.2   OECD essential elements of corporate governance for process safety [26].

TABLE 21.2 Sample Self-Assessment Questions for the CEO and Senior Leaders According to the OECD Recommended Elements of Corporate Governance for Process Safety [26]

Element	Question
Leadership and Culture: Creating an open environment	Do you and senior leaders actively work to remove any barriers to the reporting of "bad news" up the management hierarchy, and promote an open culture for communicating process safety issues (e.g., by providing direct communication routes from the shopfloor to senior leaders, or from the national board to overseas HQ)?
Risk Awareness: Broadly understanding vulnerabilities and risks	Do you and your senior leaders understand the means of prevention, control, and mitigation of major process safety hazards?
Information: Ensuring data drives process safety programs	Do you have a process safety management system (this may be integrated into a broader HSEQ management system)? [HSEQ = Health, Safety, Environment, Quality]
Competence: Assuring the organization's competence to manage the hazards of its operations	Do you have effective process safety competency requirements for all personnel with process safety impacts, including you and senior leaders?
Action: Articulating and driving active monitoring and plans	Do you and your senior leaders have accountability for the completion of corrective actions identified in significant audits, inspections, investigations, and management of change assessments, etc.?

senior corporate leaders, the full set of five to eight questions per element makes for interesting reading for anyone involved in an industry where combustible dusts are handled—as does the entire OECD document [26] itself. The commute from my residence to Dalhousie University takes about one hour by public transit. I was able to comfortably read and reflect on the OECD report [26] during one trip to work. Might you find the time to do the same?

REFERENCES

[1] Amyotte PR, Eckhoff RK. Dust explosion causation, prevention and mitigation: an overview. Journal of Chemical Health & Safety 2010;17:15–28.

[2] Hendershot DC. Process safety management—you can't get it right without a good safety culture. Process Safety Progress 2012;31:2–5.

[3] MKOPSC. A frontiers of research workshop. Process safety research agenda for the 21st century. A policy document developed by a representation of the global process safety academia (October 21–22, 2011). College Station, TX: Mary Kay O'Connor Process Safety Center, Texas A&M University System; 2012.

[4] Hopkins A. Lessons from Longford. The Esso gas plant explosion. Sydney, Australia: CCH Australia Limited; 2000.

[5] Hopkins A. Safety, culture and risk. The organizational causes of disasters. Sydney, Australia: CCH Australia Limited; 2005.

[6] Hopkins A. Failure to learn. The BP Texas City Refinery disaster. Sydney, Australia: CCH Australia Limited; 2009.

[7] Hopkins A, editor. Learning from high reliability organisations. Sydney, Australia: CCH Australia Limited; 2009.

[8] Amyotte P, Khan F, Demichela M, Piccinini N. Industrial safety culture—leading indicators and application. Hong Kong, China: Proceedings of Ninth International Probabilistic Safety Assessment and Management Conference (PSAM 9), Paper 0114; May 18–23, 2008.

[9] Earls AR. Refining the process. NFPA Journal (March/April). 2010. (Available at: http://www. nfpa.org/publicJournalDetail.asp?categoryID=1965&itemID=46404&src=NFPAJournal &cookie_test=12010; last accessed December 11, 2012.)

[10] CSB. Investigation report—sugar dust explosion and fire—Imperial Sugar Company. Report No. 2008-05-I-GA. Washington, DC: U.S. Chemical Safety and Hazard Investigation Board; 2009.

[11] Franck A, editor. Imperial Sugar CEO latest to call for OSHA combustible dust standard. Hazardous Area International 2012;6(3):5.

[12] CCPS (Center for Chemical Process Safety). Guidelines for risk based process safety. Hoboken, NJ: John Wiley & Sons, Inc; 2007.

[13] CSChE. Process safety management guide, 4th ed. Ottawa, ON: Canadian Society for Chemical Engineering; 2012. (Available at: http://www.cheminst.ca/index.php?ci_id=3638&la_id=12012; last accessed December 12, 2012.)

[14] Amyotte PR, Goraya AU, Hendershot DC, Khan FI. Incorporation of inherent safety principles in process safety management. Orlando, FL: Center for Chemical Process Safety, American Institute of Chemical Engineers, Proceedings of 21st Annual International Conference—Process Safety Challenges in a Global Economy; April 23–27, 2006, pp. 175–207.

[15] Energy Institute. Hearts and minds. London, UK: Energy Institute; 2012. (Available at: http://www.eimicrosites.org/heartsandminds/2012; last accessed December 12, 2012.)

[16] Morozzo C. Account of a violent explosion which happened in the flour-warehouse, at Turin, December the 14th, 1785; to which are added some observations on spontaneous inflammations. From the Memoirs of the Academy of Sciences of Turin (London: The Repertory of Arts and Manufactures) [Version published with foreword by N. Piccinini, Politecnico di Torino (1996)], 1795.

[17] Richard KP, Justice. The Westray story—a predictable path to disaster. Report of the Westray Mine Public Inquiry. Halifax, NS, Canada: Province of Nova Scotia; 1997.

[18] Di Menna J. Safety haven. Canadian Chemical News 2012;64(9):24–7.

[19] Amyotte PR, Oehmen AM. Application of a loss causation model to the Westray mine explosion. Process Safety and Environmental Protection 2002;80:55–9.

[20] Bird FE, Germain GL. Practical loss control leadership. Loganville, GA: Det Norske Veritas; 1996.

[21] OHS. Are all accidents preventable? OHS Canada 2000;16(5):30–6.

[22] Lian J. Raising the safety bar: an update on prevention efforts. OHS Canada 2012;28(8):20–1.

[23] McAteer JD, Beall K, Beck JA Jr, McGinley PC, Monforton C, Roberts DC, Spence B, Weise S. Upper Big Branch. The April 5, 2010, explosion: a failure of basic coal mine safety practices. Report to the Governor Governor's Independent Investigation Panel 2011. (Available at: http://nttc.edu/programs&projects/minesafety/disasterinvestigations/upperbigbranch/ UpperBigBranchReport.pdf2011; last accessed December 12, 2012.)

[24] Moure-Eraso R. CSB Chair Moure-Eraso addresses nationwide meeting of International Code Council, saluting progress on key safety issues and encouraging further code action on combustible dust hazards (October 22, 2012). Washington, DC: U.S. Chemical Safety and Hazard Investigation Board; 2012.

[25] CSB. Case study—Hoeganaes Corporation: Gallatin, TN—metal dust flash fires and hydrogen explosion. Report No. 2011-4-I-TN. Washington, DC: U.S. Chemical Safety and Hazard Investigation Board; 2011.

[26] OECD. Corporate governance for process safety. Guidance for senior leaders in high hazard industries. OECD Environment, Health and Safety. Chemical Accidents Programme 2012. (Available at: http://www.oecd.org/chemicalsafety/ riskmanagementofinstallationsandchemicals/49865614.pdf2012; last accessed December 10, 2012.)

Conclusion: Dust Explosion Realities

Twenty myths that play a role in the occurrence of dust explosions have been identified in this book on the basis of my research and practice experience over the past three decades. As noted in Chapter 1, others would likely come up with a different list; perhaps future presentation of these thoughts will help in communicating practical solutions to help resolve the dust explosion problem. Communication of the realities of dust explosions must be an ongoing concern.

In these communication efforts, it is vital that we draw on the teachings afforded us by the natural, management, and social sciences, as well as the fundamental principles of applied science. Engineering is often about choosing the "best" solution from a number of good alternatives; there will thus always be a need for interpretation and opinion when it comes to engineering practice. The choices made will be more amenable to implementation when they flow from a basis in fact.

22.1 MYTHS AND CORRESPONDING REALITIES

To conclude on a factual note, I present in Table 22.1 the realities associated with the 20 myths identified here.

22.2 WHAT DO *YOU* THINK?

Are there myths about dust explosions that you have encountered in your workplace and that have not been discussed in this book? What factual response would you provide to a colleague who holds one or more of these myths as being true?

To stimulate your thought process, I have listed below a number of myths provided to me by one of the anonymous reviewers of the proposal I submitted to the publisher for the current book. The reviewer commented that some of these might already be covered by the 20 myths I had proposed to form the basis of the book. What do you think—have you heard these beliefs expressed before—here or elsewhere? How would you respond to them?

Here is the list as originally provided by the reviewer, with much gratitude to that person:

- If it doesn't burn as a bulk solid, then it can't burn or explode as a fuel. Thus, metal dust does not explode.
- My dust has a large particle size, so an explosion with this dust is not possible. Testing of the dust is not required.
- Dust layers are not a dust explosion hazard since the dust needs to be airborne to ignite.
- Electric motors and light bulbs will not ignite dust layers since their surface temperature is not hot enough to ignite the dust layer.
- Electrostatic sparks are not energetic enough to ignite dusts.
- Dust layers need to be really thick to provide adequate dust for an explosion.

TABLE 22.1 Myths and Realities of Dust Explosions Explored in this Book

Myth (Fiction)	Reality (Fact)
Dust does not explode.	Many of the dusts handled in industry are combustible and therefore present an explosion hazard. To conclude that a given material is non-explosible requires incontrovertible evidence.
Dust explosions happen only in coal mines and grain elevators.	Dust explosions occur in a wide range of industries and industrial applications involving numerous and varied products such as coal, grain, paper, foodstuffs, metals, rubber, pharmaceuticals, plastics, textiles, etc.
A lot of dust is needed to have an explosion.	Combustible dust clouds can be generated from layers of dust deposited on surfaces in thicknesses on the scale of a millimeter (mm) or less.
Gas explosions are much worse than dust explosions.	Both gas and dust explosions have the potential to cause loss of life, personal injury, property damage, business interruption, and environmental degradation. The likelihood of occurrence and severity of consequences for a given fuel/air explosion, as well as appropriate prevention and mitigation measures, are most effectively determined by a thorough assessment of the specific material hazards and process conditions.
It's up to the testing lab to specify which particle size to test.	Selection of a dust sample for explosibility testing requires close collaboration between the test facility and plant personnel. A key selection consideration is the particle size distribution of the sample to be tested; the test facility alone cannot specify this material property.
Any amount of suppressant is better than none.	Effective suppression of a dust explosion requires that sufficient inert dust be admixed with the fuel dust. Amounts of suppressant less than that required for complete suppression can lead to higher explosion pressures than for the case of the fuel dust alone (i.e., no added suppressant).
Dusts ignite only with a high-energy ignition source.	Energetic ignition sources on the order of several thousand Joules are routinely used in closed-vessel dust explosibility testing to overcome ignitability limitations imposed by the harsh test conditions. There is clear evidence that some dusts will ignite at spark energies less than 1 mJ (milliJoule) under conditions of lower turbulence intensity.
Only dust clouds—not dust layers—will ignite.	Dust layers pose a unique hazard that is different from that presented by dust clouds. Dust layers themselves do not explode. They can, however, ignite due to self-heating or hot surfaces and smolder or cause flaming combustion.
Oxygen removal must be complete to be effective.	For a given dust, there are volume percentages of oxygen (less than the typical 21 volume % in air) for which an explosion will not be possible. The highest of these concentrations is known as the limiting oxygen concentration, or LOC.

Continued

TABLE 22.1 Myths and Realities of Dust Explosions Explored in this Book— cont'd

Myth (Fiction)	Reality (Fact)
Taking away the oxygen makes things safe.	While it is true that some alternatives are *safer* than others, nothing is *safe*. Further, an attempt to remove one hazard can, in fact, introduce another. For example, replacement of oxygen with nitrogen can eliminate a dust explosion hazard while at the same time introducing an asphyxiation hazard.
There's no problem if dust is not visible in the air.	Absence of airborne dust in work areas does not indicate the absence of a dust explosion hazard. Because explosible dust clouds are optically thick, dust explosions occur as primary events inside process units (i.e., where people are not normally present). Secondary explosions caused by initially layered dust in work areas can, however, cause significant loss.
Once airborne, a dust will quickly settle out of suspension.	A number of factors have an influence on the ease of dispersion of a dust and the tendency of the particulate matter to remain airborne once a dust cloud has been formed. These factors include particle size, shape and specific surface area, dust density and moisture content, and agglomeration processes both pre- and post-dispersion.
Mixing is mixing; there are no degrees.	The role of variable turbulence in determining the state of *mixedness* of a dust cloud is a dominant concern in understanding dust explosion likelihood and severity.
Venting is the only/ best solution to the dust explosion problem.	Dust explosion prevention and mitigation are most effectively accomplished by a hierarchical consideration of the various measures available—ranging, in decreasing order of effectiveness, from inherently safer design to engineered safety (passive and active) to procedural safety. Explosion relief venting is a passive engineered approach to explosion mitigation.
Total confinement is required to have an explosion.	The degree of confinement that will enable pressure buildup need not be complete. Partial confinement can occur as a facility design feature or can result from the actual process of explosion relief venting.
Confinement means four walls, a roof, and a floor.	Congestion in a workplace as well as the presence of turbulence-generating obstacles can facilitate flame acceleration and subsequent overpressure development. It is also possible to create a degree of confinement unintentionally by means of temporary structures.
The vocabulary of dust explosions is difficult to understand.	Dust explosion terminology, as used to describe various parameters such as maximum explosion pressure and rate of pressure rise, has a direct analogy to that used for gas explosions. It is incumbent on the management of an industrial facility to ensure that its workforce is familiar with this terminology from a process safety perspective and to the extent needed for hazard awareness in relation to specific job functions.

TABLE 22.1 Myths and Realities of Dust Explosions Explored in this Book — cont'd

Myth (Fiction)	Reality (Fact)
Dust explosion parameters are fundamental material properties.	Dust explosion parameters such as the size-normalized maximum rate of pressure rise, K_{St}, are strongly dependent on material characteristics such as particle size and experimental conditions such as turbulence intensity. Therefore, they are not intrinsic or fundamental material constants. Standardized equipment and test methodologies are available for determining these parameters; these methodologies must be followed to generate test data for use with the measures identified in dust explosion prevention and mitigation standards.
It makes sense to combine explosion parameters in a single index.	There is no single index appropriate for an assessment of the overall dust explosion risk posed by a given material. Such an assessment, as previously mentioned, requires consideration of specific material hazards and process conditions. Most material safety data sheets for combustible dusts are inadequate in terms of conveying these specific material hazards.
It won't happen to me.	Belief that a dust explosion will not happen in a given facility handling combustible powders is rooted in an inadequate safety culture. The end result of such a belief is inevitably the very thing it denies—a dust explosion.

- Since dust is difficult to keep airborne, I need to do dust housekeeping only in the immediate vicinity of the dust work area and only in directly visible areas.
- Common materials that we use safely every day, such as milk powder and sugar, cannot possibly cause a dust explosion.
- We can prevent dust explosions by eliminating ignition sources alone.
- Dust cannot collect in ventilation ducts; the airflow will keep it airborne.
- Pneumatic conveying of dust is the best and safest way to transport dust materials.
- Dust housekeeping can be easily done using air hoses and a common industrial vacuum.
- Dust housekeeping needs to be done only on a periodic basis, such as once a week.
- I cannot have a dust explosion outdoors in the open.
- I can hire an outside contractor to do the dust housekeeping and then assume the problem has been solved.
- Interpretation and application of the dust testing data is the easy part.
- Explosion vents on buildings and vessels can be directed to the most convenient location, including work areas.

- Explosion vents on buildings will protect the inhabitants from the effects of an explosion.
- Dust explosion data obtained from any experimental conditions or apparatus are adequate to design dust explosion protection systems.
- Dust explosions can be controlled by either prevention or mitigation. Only one is required for a dust explosion control program.
- There are no electrical code requirements for dusty environments.
- No rigorous definition exists to classify a location as a dusty environment. (The reviewer notes that even he or she is not sure about this one.)

In terms of refuting some of the preceding statements, you will undoubtedly have your own scientific explanations and examples. The dust explosion literature can also provide factual, authoritative statements to help in this regard. Consider the following excerpt from Glor [1] (pp. 3–4) in response to at least two of the preceding items (with specific reference to powder transfers in the pharmaceutical industry):

As previous sections have outlined, it is nigh on impossible to prevent the formation of explosive atmospheres. Additionally, the exclusion of effective ignition sources from a process is not simple and can in no way be a guaranteed measure against explosion risks. It is necessary for employers to utilize every possible precaution to prevent explosions from happening and protect both their plant and their personnel.

As a final exercise, can you assign each of the preceding myths to one or more of the elements of the explosion pentagon shown in Figure 1.3? (See Section 1.2 for examples.)

REFERENCES

[1] Glor M. A synopsis of explosion hazards during the transfer of powders into flammable solvents and explosion preventative measures. Pharmaceutical Engineering 2010;30:1–8.

Index

Note: Page numbers followed by "f" and "t" denote figures and tables, respectively.

A

ABS. *See* Acrylonitrile-butadiene-styrene
Acetone gas mixtures, explosion regimes for, 48–50, 49f
Acrylonitrile-butadiene-styrene (ABS), 84–85, 85f, 116–117
Active engineered safety, 171–172, 174t, 175f
Agglomeration process, 143, 145–146
AIChE. *See* American Institute of Chemical Engineers
ALARP. *See* As low as reasonably practicable
Alex (nano-size aluminum) dust explosion, 108t
Aluminum
  explosibility parameters, particle size influence on, 58, 58t, 60
  granulate, minimum ignition energy, 82t
  limiting oxygen concentration of, 103t
  qualitative, 224
  quantitative, 224, 225f
Aluminum dust explosion, 22f, 70–71
  risk analysis of, 224
American Institute of Chemical Engineers (AIChE)
  Center for Chemical Process Safety, 237
American Society for Testing and Materials (ASTM), 12, 58–59, 215
Argon
  magnesium with, limiting oxygen concentration for, 107t
  specific heat capacity of, 106t
As low as reasonably practicable (ALARP), 114, 115f
ASTM. *See* American Society for Testing and Materials

B

BAM oven, 80–81, 81f
BEM. *See* Bruceton Experimental Mine
Benzanthron, minimum ignition energy of, 82t
Biomass dust, 222, 222t
Bloom's Taxonomy of Educational Objectives, 177–178
Bow-tie analysis (BTA), 97, 98f
BP Texas City refinery, vapor cloud explosion at, 236

Bruceton Experimental Mine (BEM), 144–145
BTA. *See* Bow-tie analysis

C

CA. *See* Checklist analysis
Cadmium stearate, limiting oxygen concentration of, 103t
Canadian Society for Chemical Engineering (CSChE)
  process safety management of, 238, 239t
Candidate inert gases, 106–107
  specific heat capacity of, 106t
Carbon dioxide
  magnesium with, limiting oxygen concentration for, 107t
  specific heat capacity of, 106t
CCPS. *See* Center for Chemical Process Safety
CEI. *See* Chemical Exposure Index
Center for Chemical Process Safety (CCPS), 119, 237
CFD. *See* Computational fluid dynamics
Checklist analysis (CA), 224
Chemical explosion, 190
Chemical Exposure Index (CEI), 220
Chemical inhibitors, 67. *See also* Thermal inhibitors
Chemical process industries (CPI), 182, 183f
Coal dust, 222, 222t
  bituminous, 41, 44f, 53–54
  Colombian, physical properties of, 56t
  Datong, physical properties of, 56t
  and explosion pressure, 55, 55f
  housekeeping, 36
  MIC values with inertants, 69–70, 70f
  and rate of pressure rise, 55, 55f
Coal dust explosion, 18, 46
  frequency of, 25, 25f
  methane-triggered, 18, 20f, 32, 32f, 46
  pressure, influence of dolomite on, 72f
  pressure data for, 46–47, 47f
Coal pile fire, evolution of, 93f
Coke dust/fly ash explosion, 13, 14f
Cold petrol, 14
Collective mindfulness, 237, 244
Combustible dust, defined, 10–11

Combustion
  heterogeneous, 40–41
  homogeneous, 40–41
  spontaneous, 92, 96–99
Completely confined explosion, 184, 190
Computational fluid dynamics (CFD), 61–62,
    62f, 209
Concentration gradients, 161–163
Confinement, 167–180, 189–198
  complete, 190
  degree of, 182–184
  partial, 190
  total, 181–188
Congestion, 190–193
Controls, hierarchy of, 35, 171–172
Corn dust, explosion data for, 221, 222t
Corn starch explosion, 66f
CPI. *See* Chemical process industries
CSB. *See* U.S. Chemical Safety Board
CSChE. *See* Canadian Society for Chemical
    Engineering
Cube-root law, 211
Culture
  flexible, 237
  just, 237
  learning, 237
  reporting, 237
  safety, 236–238, 244

D
DDT. *See* Deflagration-to-detonation transition
DeBruce grain elevator explosion, 130
Deepwater Horizon disaster, 236
Deflagration, 190
Deflagration-to-detonation transition (DDT),
    190–191
Degree of confinement, 182–184
DESC. *See* Dust Explosion Simulation Code
Detonation, 190–191
Dolomite, 36, 69, 71–72
  influence on explosion pressure of coal dust,
    72f
Domino effect, 132–133, 135
Dust, defined, 10–11
Dust cloud generation, potential for, 33f
Dust cloud ignition by low-energy sources,
    83–87
Dust density, 143
Dust dispersion, 143–144
Dust explosions
  chemical, 190
  classification scheme for, 183f
  completely confined, 184

defined, 182
determination of, 12
modeling, 61–62
parameters, 42, 43t
partially confined, 184
phenolic resin, 22f
physical, 190
prevention of, 172–176, 173f, 174t
primary, 128–132, 129f
reality of, 249–254
relief venting, 184–186
risk assessment, 223–226
risk management, 223–226
risk matrix for, 226f
secondary, 128–132, 129f
severity, 220, 222
terminology, 200–203
triggers, 78
unconfined, 184
vocabulary of, 199–206
Dust Explosion Simulation Code (DESC),
    61–62, 62f, 224
Dust layer fires, 93–95
Dust layer ignition, 92–93
  reality of, 95–96
Dust moisture content, 143
Dustiness, 140–143
  defined, 140–141
  factors influencing, 141, 142t

E
Epoxy coating powder, minimum ignition
    energy of, 82t
Epoxy-polyester coating powder, minimum
    ignition energy of, 82t
Escalation vector, 135–136, 136t
ETA. *See* Event tree analysis
Event tree analysis (ETA), 97–98, 98f
Explosibility index, 220
Explosible non-explosible dust, 12–13
Explosion-generated turbulence.
    See Post-ignition turbulence

F
F&EI. *See* Fire & Explosion Index
Failure modes and effects analysis (FMEA), 224
Fault tree analysis (FTA), 97–98, 98f, 224, 225f
Fibers, 14–16
Fire & Explosion Index (F&EI), 220
Firedamp, 18
Flameless venting, 185, 185f
Flexible culture, 237
Flocculent-materials dust, 146–149

Flock, minimum ignition energy of, 82t
Flour, 18
Fly ash explosion, 12–13, 46
  coke dust/fly ash explosion, 13, 14f
  explosible mixture, volatile percentage
    required for, 13, 13f
FM Global, 21–22
FMEA. See Failure modes and effects analysis
FTA. See Fault tree analysis
Fuel carryover, 12–13
Fuel with oxidizing atmosphere, mixing of,
  127–138

G

Gallatin, TN, Hoeganaes plant, iron dust
  incident at, 130–132, 186
Gas explosion, 10, 39–50
  analogies, 203–204
  hybrid mixtures, 45–46
  likelihood of occurrence and prevention, 42–44
  severity of consequences and mitigation, 45
Gasoline. See Cold petrol
Germany, dust explosions frequency, 24–25, 24f
  in industries handling wood and coal dusts
    in, 25, 25f
Grain elevator explosion, 18, 19f
  human and financial impacts, 21t

H

Hammermill, 25–26, 26f
Hazard, 40, 42, 117–118
Hazard and operability analysis (HAZOP), 224
HAZOP. See Hazard and operability analysis
Heterogeneous combustion, 40–41
Hierarchy of controls, 35, 171–172
High-rate discharge (HRD) suppressant
  systems, 67–68, 68f
Homogeneous combustion, 40–41
Hot surfaces, dust layer ignition by, 92–93
Hot Wire Anemometry (HWA), 163
Housekeeping, 34–36
HRD. See High-rate discharge suppressant
  systems
HSE. See UK Health and Safety Executive
Human body, oxygen deficiency effects on,
  102, 102t
HWA. See Hot Wire Anemometry
Hybrid mixtures explosion, 45–46

I

Ignition sensitivity, 220, 222
Ignition sources, 77–100
  industrial, 78–79

low-energy sources, dust cloud ignition by,
  83–87
  prevention of, 42–44, 44f
  reality of, 87–88
  standardized dust explosibility testing of,
    79–83
Imperial Sugar Company, Port Wentworth, GA,
  explosion at, 196, 237
Incident pyramid, 201–202, 201f
Inductance, influence on minimum ignition
  energy, 89, 89f
Industrial ignition sources, 78–79
Inert gas, candidate, 106–107
Inerting, 67–69, 67f
  partial, 108–109
  total, 107
Inherently safer design (ISD), 115, 134,
  169–176, 238
  minimization, 169t, 170, 174t, 177
  moderation, 169t, 170, 174t, 177
  simplification, 169t, 170–171, 174t, 177,
    178f
  substitution, 169t, 170, 174t, 177
Iron dust combustibility, 130–132, 132f
Iron dust explosion, 6, 7f
ISD. See Inherently safer design

J

Japan
  1952–1995 dust explosions, analysis of, 27t
Just culture, 237

K

Kleen Energy, Middletown, gas blow at, 117,
  118f
Knock-on effect. See Domino effect

L

Lake Lynn Experimental Mine (LLEM),
  144–145
Laminar burning velocity, 208–209
Laser Doppler Anemometry (LDA), 158
Layer ignition temperature (LIT), 93, 94f, 203
LDA. See Laser Doppler Anemometry
Learning culture, 237
LFL. See Lower flammability limit
Likelihood of occurrence, 220
Limestone, 12, 36, 69, 71–73
  as inertant for coal dust, 69–70, 70f
Limiting oxygen concentration (LOC),
  103–106, 103t, 107t
LIT. See Layer ignition temperature
LLEM. See Lake Lynn Experimental Mine

LOC. *See* Limiting oxygen concentration
Lost-time injuries (LTIs), 202
Low-energy sources, dust cloud ignition by, 83–87
Lower flammability limit (LFL), 45–46, 203
LTIs. *See* Lost-time injuries
Lycopodium
  minimum ignition energy, 82t
  particle size, 53

M

Magnesium dust explosibility parameters, particle size influence on, 58
Magnesium granulate, minimum ignition energy of, 82t
Major Hazard Incident Data Service (MHIDAS), 132–133, 135
Management of change (MOC), 119–120, 120f
MAP. *See* Monoammonium phosphate
Marginally explosible dusts, 214
Material safety data sheets (MSDSs), 13, 226–229
Maximum explosion pressure
  for polyethylene and polyethylene/hydrocarbon gas hybrid mixture explosions, 45f, 57, 57f
Maximum rate of pressure rise ($(dP/dt)_{max}$), 210–211
MEC. *See* Minimum explosible concentration
Methane
  explosion pressure data for, 46–47, 47f
  -triggered coal dust explosion, 18, 20f, 32, 32f, 46
MHIDAS. *See* Major Hazard Incident Data Service
MIC. *See* Minimum inerting concentration
MIE. *See* Minimum ignition energy
MIKE 3 apparatus, 79–82, 80f, 86–87, 159
Mindful organizing, 244
Minimization principle, inherently safer design, 169t, 170, 174t, 177
Minimum explosible concentration (MEC), 33, 45–46, 104, 203
  particle size influence on, 57–58
Minimum ignition energy (MIE), 81–82, 82t, 159, 205, 221
  data for niacin (nicotinic acid), 44, 44f
  defined, 81
  for dusts, 83, 85
  inductance influence on, 89, 89f
  particle size influence on, 57–58
Minimum ignition temperature (MIT), 79, 86, 86t, 205
  particle size influence on, 57–58

Minimum inerting concentration (MIC), 69–70
  for coal dust with inertants, 69–70, 70f
MIT. *See* Minimum ignition temperature
Mitigation, 172, 173f
  hierarchical view of, 172–176, 174t
Mixing, 127–166
MOC. *See* Management of change
Moderation principle, inherently safer design, 169t, 170, 174t, 177
Monoammonium phosphate (MAP), 69, 73
  as inertant for coal dust, 69–70, 70f
MSDSs. *See* Material safety data sheets
Myths of dust explosion, 4, 250, 251t–253t

N

Nano-dust ignitability, 86–87, 86t
Nano-materials dust, 145–146
Narrow size distribution dusts, 52–54
NFPA. *See* U.S. National Fire Protection Association
Niacin (nicotinic acid)
  explosion regimes for mixtures of, 48–50, 49f
  MIE data for, 44, 44f
NIOSH. *See* U.S. National Institute for Occupational Safety and Health
Nitrogen
  magnesium with, limiting oxygen concentration for, 107t
  specific heat capacity of, 106t
Normalization of deviation, 94–95
Nylon fibers, 14–16

O

Obstacle-generated turbulence, 190–193, 192f
Occupational safety
  defined, 201
  distinguished from process safety, 201–203
OECD. *See* Organization for Economic Co-operation and Development
Olive oil, 14
Organic dusts, 41
Organic pigment, limiting oxygen concentration of, 103t
Organic stabilizer, minimum ignition energy of, 82t
Organization for Economic Co-operation and Development (OECD)
  corporate governance for process safety, 244, 245f, 245t
OSHA. *See* U.S. Occupational Safety and Health Administration

Overdriving, 213–214
Oxidant component, of explosion prevention, 113–126
Oxygen
deficiency, effects on human body, 102, 102t
removal, 101–112

P

Partial confinement, 190
Partial inerting, 108–109
Partially confined explosion, 184
Particle agglomeration, 143, 145–146
Particle shape, 143
Particle size, 51–64, 143
distribution, 52–56
effects on explosibility parameters, 56–59
Particle specific surface area, 143
Passive engineered safety, 171–172, 174t, 175f
Pea flour, limiting oxygen concentration of, 103t
Pentagon, 3–4, 3f, 199–248
Pharmaceuticals, hybrid mixture explosions in, 46
Phenolic resin dust explosion, 22f
Physical explosion, 190
Pittsburgh Research Laboratory (PRL), 144–145
Plan, do, check, and act cycle, 238
Polyamide, 15, 15f
coating powder, minimum ignition energy of, 82t
polyamide 6.6, explosion sequence for, 147, 150f–151f
Polydispersity, 60
Polyester coating powder, minimum ignition energy of, 82t
Polyethylene, 11, 11f–12f, 40–41, 200
high-density, limiting oxygen concentration of, 103t
material safety data sheets, 227, 228t
particle size, 52, 54f
Polyethylene dust explosion, 21f
DESC simulations for, 61–62, 61f
maximum explosion pressure data for, 45f, 57, 57f, 61
pressure data for, 46–47, 47f
Polystyrene-acrylonitrile, 84
Polyethylene/hydrocarbon gas hybrid mixture explosion, maximum explosion pressure data for, 45f, 57, 57f
Polyurethane coating powder, minimum ignition energy of, 82t
Pomace oil, 14
Post-ignition turbulence, 159, 161, 191–193, 195

Preferential lifting, 143–145
Pre-ignition turbulence, 158–159
Pressure-piling phenomenon, 191
Prevention of dust explosion, 172, 173f
hierarchical view of, 172–176, 174t
Primary dust explosion, 128–132, 129f
Priority of controls. See Hierarchy of controls
PRL. See Pittsburgh Research Laboratory
Procedural safety, 171–172, 174t, 175f
Process safety, 176–179, 200–203
defined, 200
distinguished from occupational safety, 201–203
OECD corporate governance for, 244, 245f, 245t
Process safety management (PSM), 119
defined, 238
risk-based approach to, 238
system of Canadian Society for Chemical Engineering, 239t
PSM. See Process safety management
Pulverizing units, 25–26, 26f

Q

Qualitative risk assessment, 224
Quantitative risk assessment, 224, 225f
Quiescent dust cloud, 208–209

R

Reality of dust explosion, 14–15, 26–27, 36, 46–47, 60, 72, 87–88, 95–96, 107, 120–123, 133–135, 149–150, 163–164, 172–176, 186, 195, 204–205, 215, 227, 242–243, 249–254, 251t–253t
REMBE®, 115–117
Reporting culture, 237
Right to know, 204
Risk, 40, 42
assessment, 223–226
-awareness, 237
management, 223–226
matrix, for dust explosions, 224–226, 226f
score, 226
Rock dust, 36, 69
influence on explosion pressure of coal dust, 72f

S

Safer alternative, 114–117
Safety
active engineered, 171–172, 174t, 175f
culture, 236–238, 244

Safety (*Continued*)
  decision hierarchy. *See* Hierarchy of controls
  defined, 237
  management systems, 238–240, 239t
  measurement, 237
  occupational, 201–203
  passive engineered, 171–172, 174t, 175f
  procedural, 171–172, 174t, 175f
  process, 176–179, 200–203, 244, 245f, 245t
SBC. *See* Sodium bicarbonate
Scanning electron microscope (SEM)
  bituminous coal dust, 41, 41f
  polyamide, 15, 15f
  polyethylene, 11f–12f
Secondary dust explosion, 128–132, 129f
SEEP. *See* Suppressant Enhanced Explosion
    Parameter
Self-heating, dust layer ignition by, 92–93
Self-ignition, 92
SEM. *See* Scanning electron microscope
Simplification principle, inherently safer
    design, 169t, 170–171, 174t, 177, 178f
Siwek 20-L apparatus, 121–123, 121f, 159
  dust explosion parameters testing using,
    215–217, 216t
Size-normalized maximum rate of pressure
    (K_{St}), 210–213, 212t
  for polyethylene and polyethylene/
    hydrocarbon gas hybrid mixture
    explosions, 57, 57f, 61–62, 61f
  of wood dust, particle size effect, 58, 58t
Sodium azide, 182
Sodium bicarbonate (SBC), 69–71, 73
  as inertant for coal dust, 69–70, 70f
Spontaneous combustion, 92, 96–99
Spray drier plant
  directed graph representation for, 74f
  with explosion protection, 74f
Standardized dust explosibility testing, 79–83,
    213–215
Stone dust, 69. *See also* Rock dust
Substitution principle, inherently safer design,
    169t, 170, 174t, 177
Sugar dust explosion, 23f, 32–33, 33f
Sulfur, limiting oxygen concentration of, 103t
Suppressant Enhanced Explosion Parameter
    (SEEP), 70–72, 71f
Suppressants, 65–76
  inerting and suppression, 67–69, 67f
  minimum inerting concentration, 69–70
  Suppressant Enhanced Explosion Parameter,
    70–72, 71f

T
Temporary enclosures, 194–195
Thermal inhibitors, 67, 71–72
Thermal runaway, 97f
Thickness of dust layer, 33–34, 33f
  housekeeping, 34–36
TIC. *See* Total incombustible content
Total incombustible content (TIC), 144–145
Total inerting, 107
Turbulence, 156–158
  defined, 156
  and dust clouds concentration gradients, link
    between, 161–163
  influence on drug explosion, 158–161
  intensity, 163
  obstacle-generated, 190–193, 192f
  post-ignition, 159, 161, 191–193, 195
  pre-ignition, 158–159
  scale, 164
Turbulent burning velocity, 209

U
UK Health and Safety Executive (HSE), 237,
    239–240
Unconfined explosion, 184
Unconfined vapor cloud explosion, 40
Union Carbide plant, Hahnville, LA, nitrogen
    asphyxiation at, 194
United States, dust explosions in
  during first decade of the 21st century,
    human impact of, 23t
  frequency of, 24f
U.S. Bureau of Mines (USBM), 144–145, 230
  indices, 220–223
U.S. Chemical Safety and Hazard Investigation
    Board, 6, 19, 21–22, 24, 35–36, 118,
    130–132, 134–135, 186–187, 194, 196, 227
U.S. Chemical Safety Board (CSB). *See* U.S.
    Chemical Safety and Hazard Investigation
    Board
U.S. National Fire Protection Association,
    10–11
U.S. National Fire Protection Association
    (NFPA), 34
U.S. National Institute for Occupational Safety
    and Health (NIOSH), 144–145, 220
U.S. Occupational Safety and Health
    Administration (OSHA), 200

V
Vacuum, for dust collection, 35f
Vapor cloud explosion, 236

Venting, 167–180
  explosion relief, 184–186
  flameless, 185, 185f
Vocabulary of dust explosions, 199–206

W

West Pharmaceutical Services
  explosion damage at, 134, 135f
  two-story rubber compounding process at, 134, 134f
Westray coal mine explosion, 32–33, 32f, 36, 130, 240–242
Westwego, LA, grain elevator disaster at, 133
What-if analysis (WI), 224
Wheat flour, limiting oxygen concentration of, 103t

WHMIS. *See* Workplace Hazardous Materials Information System
WI. *See* What-if analysis
Wide size distribution dusts, 52–54
Wood
  dust explosions frequency in industries handling, 25, 25f
  particle size, 52, 53f, 58, 58t–59t
Workplace Hazardous Materials Information System (WHMIS), 226

Z

Zinc Stearate material safety data sheets, 229–230, 230t–231t

Printed in the United States
By Bookmasters